PROCEEDINGS OF THE 30TH INTERNATIONAL GEOLOGICAL CONGRESS
VOLUME 18 PART B

GEOLOGY OF FOSSIL FUELS - COAL

PROCEEDINGS OF THE 30TH INTERNATIONAL GEOLOGICAL CONGRESS

VOLUME 1 : ORIGIN AND HISTORY OF THE EARTH
VOLUME 2 : GEOSCIENCES AND HUMAN SURVIVAL, ENVIRONMENT, AND NATURAL HAZARDS
VOLUME 3 : GLOBAL CHANGES AND FUTURE ENVIRONMENT
VOLUME 4 : STRUCTURE OF THE LITHOSPHERE AND DEEP PROCESSES
VOLUME 5 : CONTEMPORARY LITHOSPHERIC MOTION / SEISMIC GEOLOGY
VOLUME 6 : GLOBAL TECTONIC ZONES / SUPERCONTINENT FORMATION AND DISPOSAL
VOLUME 7 : OROGENIC BELTS / GEOLOGICAL MAPPING
VOLUME 8 : BASIN ANALYSIS / GLOBAL SEDIMENTARY GEOLOGY / SEDIMENTOLOGY
VOLUME 9 : ENERGY AND MINERAL RESOURCES FOR THE 21ST CENTURY / GEOLOGY OF MINERAL DEPOSITS / MINERAL ECONOMICS
VOLUME 10 : NEW TECHNOLOGY FOR GEOSCIENCES
VOLUME 11 : STRATIGRAPHY
VOLUME 12 : PALEONTOLOGY AND HISTORICAL GEOLOGY
VOLUME 13 : MARINE GEOLOGY AND PALAEOCEANOGRAPHY
VOLUME 14 : STRUCTURAL GEOLOGY AND GEOMECHANICS
VOLUME 15 : IGNEOUS PETROLOGY
VOLUME 16 : MINERALOGY
VOLUME 17 : PRECAMBRIAN GEOLOGY AND METAMORPHIC PETROLOGY
VOLUME 18.A : GEOLOGY OF FOSSIL FUELS - OIL AND GAS
VOLUME 18.B : GEOLOGY OF FOSSIL FUELS - COAL
VOLUME 19 : GEOCHEMISTRY
VOLUME 20 : GEOPHYSICS
VOLUME 21 : QUATERNARY GEOLOGY
VOLUME 22 : HYDROGEOLOGY
VOLUME 23 : ENGINEERING GEOLOGY
VOLUME 24 : ENVIRONMENTAL GEOLOGY
VOLUME 25 : MATHEMATICAL GEOLOGY AND GEOINFORMATICS
VOLUME 26 : COMPARATIVE PLANETOLOGY / GEOLOGICAL EDUCATION / HISTORY OF GEOSCIENCES

PROCEEDINGS OF THE
30TH INTERNATIONAL GEOLOGICAL CONGRESS

BEIJING, CHINA, 4 - 14 AUGUST 1996

VOLUME 18 *PART B*

GEOLOGY OF FOSSIL FUELS - COAL

EDITOR:
YANG QI
CHINA UNIVERSITY OF GEOSCIENCES, BEIJING, CHINA

CRC Press
Taylor & Francis Group
Boca Raton London New York

CRC Press is an imprint of the
Taylor & Francis Group, an **informa** business

CONTENTS

The applications of high-resolution sequence stratigraphy to paralic
and terrestrial coal-bearing strata:
Two case studies from the Western North China Paleozoic Basin
and the Tulufan-Hami Jurassic Basin
Li Baofang, Wen Xianduan, Kang Xidong and Li Guidong   1

Petrology and depositional environment of Early Jurassic coal,
Western Australia
K.K. Sappal and N. Suwarna   21

Depositional evolution and coal accumulation of Ordos Basin
Wang Shuangming, Lu Daosheng and Zhang Yuping   33

Geologic factors affecting the abundance, distribution, and
speciation of sulfur in coals
C.-L. Chou   47

Multistage metamorphic evolution and superimposed metamorphism
through multithermo-sources in Chinese coal
*Yang Qi, Wu Chonglong, Tang Dazhen, Kang Xidong and
Liu Dameng*   59

Variations in coal rank parameters with depth correlated with
Variscan compressional deformation in the South Wales coalfield
R. Gayer and R. Fowler   77

Coalification jumps, stages and mechanism of high-rank coals in China
Qin Yong and Jiang Bo   99

Advances of the exploration and research of oil from coal in China
Huang Difan and Qin Kuangzong   123

Study on Jurassic coal and carbonaceous mudstone as oil source
rocks in Tuha Basin, North-Western China
*Jin Kuili, Yao Suping, Wei Hui, Tang Yaogang, Fang Jiahu
and Hao Duohu*   135

Reaction kinetics of coalification in the Ordos Basin, China
Liu Dameng, Yang Qi and Tang Dazhen   147

Proc. 30th Int'l Geol. Congr., Vol. 18, Part B, pp. 1-19
Yang Qi (Ed.)
© VSP 1997

The Applications of High-Resolution Sequence Stratigraphy to Paralic and Terrestrial Coal-Bearing Strata : Two Case Studies from the Western North China Paleozoic Basin and the Tulufan-Hami Jurassic Basin

LI BAOFANG, WEN XIANDUAN, KANG XIDONG, LI GUIDONG
Department of Energy Geology, China University of Geosciences
100083 Beijing, P. R. China

Abstract

Through the studies of high-resolution sequence stratigraphy on paralic Paleozoic coal measures of the western North China cratonic basin and terrestrial Jurassic coal measures of the Tulufan-Hami foreland basin, it is recognized that the depositional environment, coalbed thickness and maceral composition variations are mainly controlled by the relative sea/lake level changes. The main coal seam of Early Permian occurs on the bottom of the retrogradational sequence set of P_1^1 , which onlaps from the south(basinward) to the north(landward), and overlays the erosional surface of the Upper Carboniferous (2nd order sequence boundary). Along with the marine limestone beds thinning and pinching out to the north, coalbeds of four sequences combine northward into a very thick coal seam. There are no relationships between coal thickness variations and their underlying sandbody distribution because the depositional process is not continuous. Vitrinite, especially desmocollinite, of this coal seam, decreases landward to the north, and the inertinite and kaolinite interbeds increase. In the Tulufan-Hami basin vitrinite-rich and thicker coalbeds occur on the top of the retrogradational sequence, which develops during the lake level of the basin rising. Coalbeds developed in TST or retrogradational sequence sets are distributed more extensively and rich in vitrinite (especially desmocollinite), with better hydrocarbon potential than those of the coalbeds developed in HST or progradational sequence sets. Coalbeds of TST maybe is a new coal depositional model, and the previous coal depositional models [5,6] are more suitable to the coalbeds developing in HST.

Keywords: high-resolution sequence stratigraphy, western North China, Tulufan-Hami, coal depositional model

INTRODUCTION

Sequence stratigraphy analysis has very well been applied to the shoreline and shallow marine deposits, especially in the passive continental margin [9,11,12]. It can be used to resolve the detailed correlations between successions in a relatively short contemporaneous geological time interval and to find out the characteristics of the depositional responses which are controlled by the relative sea level changes. According to the principles of sequence stratigraphy, the relative sea level changes are related to the tectonic subsidence, global sea level change, sediments supply and climates[11,12].

Many paleobiology, coal geology and sedimentology research works have been done in the western North China Paleozoic Coal Basin [2,3,8,13] and the Tulufan-Hami Jurassic Coal Basin[7], but some arguments still exist such as the biostratigraphy division, the sequence stratigraphy division, the regional correlation of coal-bearing strata, the depositional environments and the regularities of coal thickness and coal quality etc. Through the research of high-resolution sequence stratigraphy, some of the above problems probably will have a better interpretation.

METHOD

These two studies are based on the outcrop investigation and drilling geophysical log analysis in the the Hedong Coalfield of the western North China Coal Basin and the Tulufan-Hami Coal Basin in northwestern China. In the Hedong Coalfield from the south to the north ,18 sections (1:200) are measured along the east side of the Yellow River (Fig. 1), and detailed works have been done mainly in the middle area of Sanjiao-Liulin district. Cores from 5 drilling boreholes are described and sampled. Several tens of drilling borehole profiles and 5 drilling geophysical logs are collected. In the Tulufan-Hami Basin 5 outcrop sections are measured, some cores are examined and sampled, and drilling geophysical logs are collected. According to the collected paleobiology and chronology data, the sequence stratigraphic units are divided and depositional cross sections are made, which shows the detailed correlation between the measured outcrop sections and the drilling borehole sections. From these works the conclusions are obtained.

STRATIGRAPHY FRAMEWORK IN WESTERN NORTH CHINA

Research works on stratigraphy of the Paleozoic coal-bearing strata in the North China cratonic basin have started for about a hundred years . The latest biostratigraphic and chronostratigraphic works have been done by the General Coal Geology and Exploration Bureau [8,13]. The stratigraphic unit division is showed in Fig. 2.

Lithostratigraphic Unit
According to the lithofacies association, the Permo-Carboniferous coal-bearing strata in North China are divided into Benxi, Taiyuan, Shanxi and Shihezi Formations. The Lower and Upper Shihezi Formations contain workable coalbeds in southern North China, but not in this area.. The Benxi Formation and Lower Member of Taiyuan Formation consist mainly of shoreline and shallow marine clastic deposits, with 3-5 thin carbonate beds. The Middle and Upper Member of Taiyuan Formation consist of bioclastic limestones and intercalated with some sandstones, mudstones and coalbeds, represent carbonate ramp mixed with terristrial clastic deposits. Shanxi Formation consists of deltaic clastic deposits and with mineable coalbeds intercalated. Gray green coloured clastic deposits of Lower Shihezi Formation are developed on fluvial plain. Lithofacies characteristics of the Paleozoic coal-bearing strata in the Hedong Coalfield are very similar to the representative section of the West Hill of Taiyuan, where it is not far from the middle of the research area.

Chrono-biostratigraphic Unit
Recent biostratigraphy works in Hedong Coalfield have been done by Zeng Xuelu and Yang Guanxiu [2,3], Li Yunlan [8] and Wang Zengyin [13]. Three *Fusulinids Zone* were identified in Benxi and Taiyuan Formations : *Fusulina-Fusulinella Zone*   occurs in the limestones of Benxi Formation, *Triticites zone and Montiparus Zone* occur in the Lower Member of Taiyuan Formation, and *Pseudoschwagerina Zone* in the Middle and Upper Member of Taiyuan Formation. Because there are no marine beds in Shanxi Formation, the plant fossils *Emplectoptaris trangularis--Taeniopteris muctonata--Lobatannularia sinensis Assemblage* occur in that stage.

In China the *Pseudoschwagerina zone* used to represent the latest fossils of Upper Carboniferous. But in other countries these fossils are commonly considered as the earliest fossils zone occurring in the beginning of Early Permian. In order to correlate with other countries many of the Chinese geologists agree to put the C-P boundary under the marine beds which containing the *Pseudoschwagerina Zone* [Huang Jiqing et al., 13]. In this research we divide the C-P boundary between the main coalbed of Taiyuan Formation and its underclay,   which is an exposed paleosol

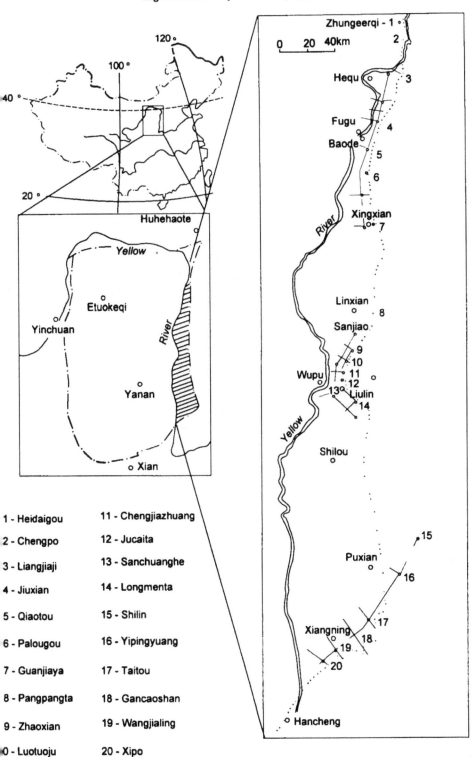

1 - Heidaigou
2 - Chengpo
3 - Liangjiaji
4 - Jiuxian
5 - Qiaotou
6 - Palougou
7 - Guanjiaya
8 - Pangpangta
9 - Zhaoxian
10 - Luotuoju

11 - Chengjiazhuang
12 - Jucaita
13 - Sanchuanghe
14 - Longmenta
15 - Shilin
16 - Yipingyuang
17 - Taitou
18 - Gancaoshan
19 - Wangjialing
20 - Xipo

Figure 1. Index map of Hedong Coalfield, shows the distribution of the outcrops, the locations of the measured

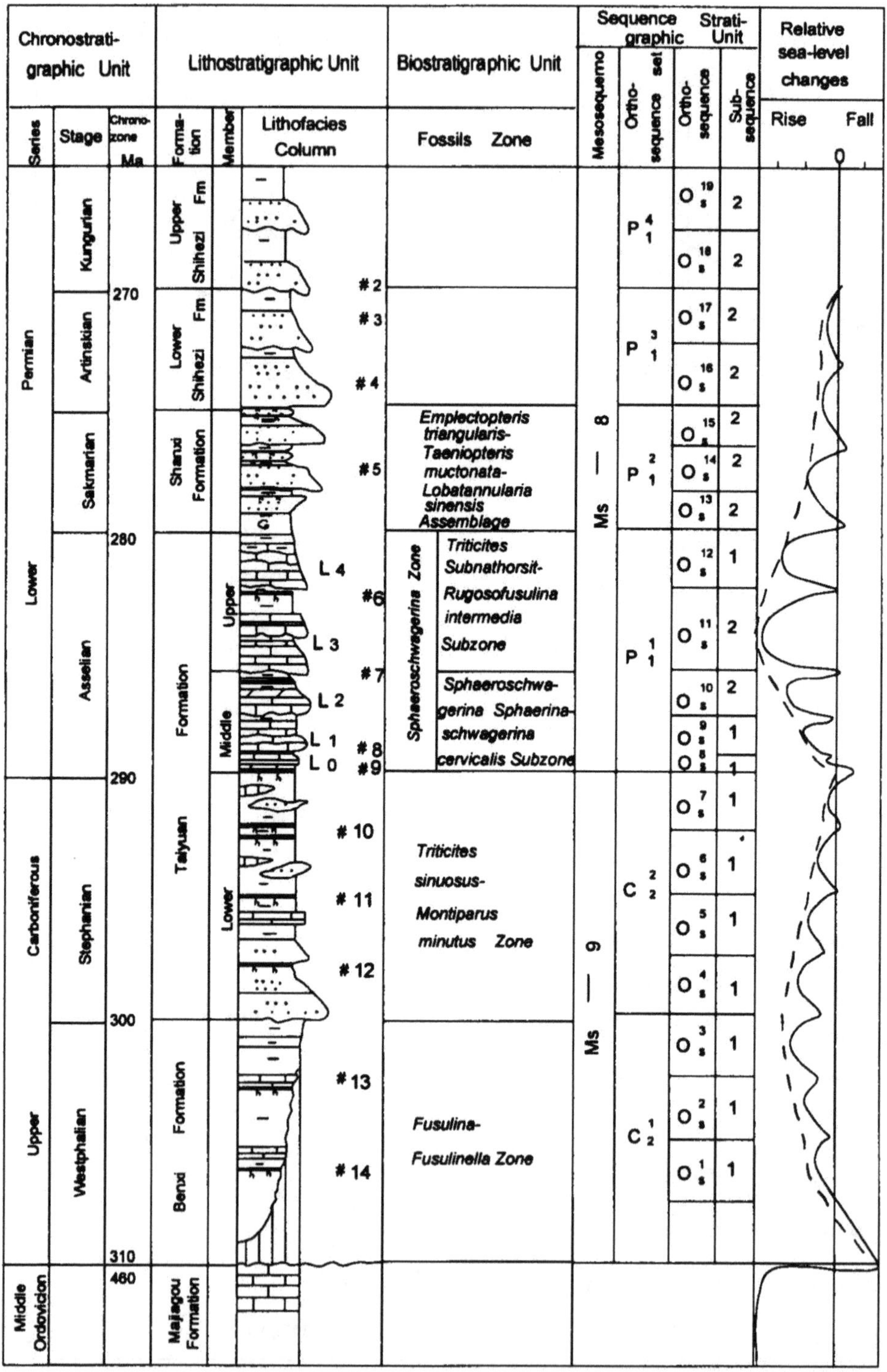

Chronostratigraphic Unit			Lithostratigraphic Unit			Biostratigraphic Unit	Sequence graphic		Stratigraphic Unit		Relative sea-level changes	
Series	Stage	Chronozone Ma	Formation	Member	Lithofacies Column	Fossils Zone	Mesosequence	Orthosequence set	Orthosequence	Subsequence	Rise	Fall
Permian	Kungurian	270	Upper Shihezi Fm				Ms — 8	P_1^4	O_s^{19}	2		
					#2				O_s^{18}	2		
	Artinskian		Lower Shihezi Fm		#3			P_1^3	O_s^{17}	2		
					#4				O_s^{16}	2		
Lower	Sakmarian	280	Shanxi Formation		#5	Emplectopteris triangularis-Taeniopteris muctonata-Lobatannularia sinensis Assemblage		P_1^2	O_s^{15}	2		
									O_s^{14}	2		
									O_s^{13}	2		
	Asselian		Taiyuan Formation	Upper / L4 / L3	#6 / #7	Triticites Subnathorsit-Rugosofusulina intermedia Subzone		P_1^1	O_s^{12}	1		
									O_s^{11}	2		
				Middle / L2 / L1 / L0	#8 / #9	Sphaeroschwagerina Sphaerina-schwagerina cervicalis Subzone			O_s^{10}	2		
		290							O_s^9	1		
									O_s^8	1		
Carboniferous	Stephanian			Lower	#10 / #11 / #12	Triticites sinuosus-Montiparus minutus Zone	Ms — 9	C_2^2	O_s^7	1		
									O_s^6	1		
									O_s^5	1		
		300							O_s^4	1		
Upper	Westphalian		Benxi Formation		#13				O_s^3	1		
					#14	Fusulina-Fusulinella Zone		C_2^1	O_s^2	1		
		310							O_s^1	1		
Middle Ordovician		460	Majiagou Formation									

Figure 2. Column section shows the chrono - biostratigraphic unit, the lithostratigraphic unit and the relative sea - level cycle curve in the western North China Paleozoic Coal Basin. Data of fossils zone are from Liyunlan[8].

of the top of Upper Carboniferous, and its overlying coalbed represents the beginning of Early Permian.

The Paleozoic coal-bearing strata overlay the Middle Ordovician Majiagou Formation with unconformity contact. According to the IUGS 1989 Global Stratigraphic Chart (by J. W. Cowie and M. G. Bassett) and magnetic stratigraphy work which has been done by Zhu Hong [13], the Benxi Formation is developed from 307.1-300 Ma, belongs to Westphalian, and the Lower Member of Taiyuan Formation is developed from 300- 290 Ma, belongs to Stephanian, Upper Carboniferous. The Middle and Upper Member of Taiyuan Formation are formed from 290-280, belong to Asselian of Early Permian. Shanxi Formation, possibly including Lower Shihezi Formation in the research area, is formed from about 280-270, belongs to Sakmarian and Artinskian of Early Permian.

In the research area the *Fusulina-Fusulinella Zone* is only distributed in the north and middle part of the Hedong Coalfield. Benxi Formation becomes thinning and overlaps southward. The *Triticites Zone* occurs more extensive, but the carbonate rocks are not continuous laterally. The *Pseudoschwagerina Zone* distributed almost all over in the Shanxi province, but not in the southern Inner Mongolia. A special kind of *Triticites* occur on the bottom limestone (L_0) in southern Hedong Coalfield, represents the transitional fossils from the *Triticites Zone* to *Pseudoschwagerina Zone*, its geological time significance should be the very beginning of Early Permian. There are some differences of the fossils in the *Pseudoschwagerina Zone*, that is the genus of the *Sphaeroschwagerina* which is not common in the lower three limestones (L_0, L_1, L_2), but very develop in the upper two limestones (L_3, L_4).

Sequence Stratigraphic Unit

As the lithostratigraphic unit is diachronic in a large area, sequence stratigraphy unit division is more scientific. Sequence stratigraphy studies the sequence of strata formed in an isochronic stratigraphy framework, the geological time interval of which can be divided as short as an epoch or stage. Sequence is bounded by uncomformities or its response conformity surfaces, forming in subaqueous environments[9,11]. Because of the cyclic evolution in the geological history, sequence can be divided into different hierarchy. In this research we followed the scheme of hierarchy for sequence stratigraphy presented by Wang Hongzhen and Shi Xiaoying [14]. According to their classification the Upper Carboniferous in China belongs to the 2nd order of sequence cycle named Mesosequence (Ms) 9, and the Permian belongs to Ms-8. The time interval of a mesosequence is about 30-40 Ma. Orthosequence set is a smaller 2nd order cycle, the time interval of which is about 9 -12 Ma. The 3rd cycle of sequence is named Orthosequence (Os), and the time interval is about 2-5 Ma, equivalent to the sequence which is defined by Mitchum et al. [9]. The 4th cycle of sequence calls Subsequence (Ss), its time interval is about 0.1-0.4 Ma, equivalent to parasequence which is commonly used in sequence stratigraphy [9] and is response to the longer Milankovitch cycle [14].

As the Paleozoic coal measures of the North China are developed in a huge cratonic basin, the depositional rate is quite slow and discontinuous. The whole thickness of the strata is only 120-150 m, but the time interval is about 35-40 Ma (from 310-275 or 270). The sequence stratigraphic units of the Benxi to Lower Shihezi Formations are divided into 2 Mesosequences, including 5 Orthosequence sets and 17 Orthosequences in the study area.

CORRELATIONS OF SEQUENCE STRATIGRAPHIC UNITS

Regional sequence stratigraphic unit correlation between the sections is a basic and important work.. In the study area the outcrops of the coal measures distribute more than 400 km from the north to the south, 18 measured sections are showed on the cross section and need to correlate in isochrone stratigraphy framework.

Basis of Correlation

1. Paleontological fossils assemblage; it is the most important basis for the 1st and 2nd cycle sequence stratigraphic analysis, but can not resolve the detailed correlations inside in an 3rd or 4th cycle sequence by only using it.
2. Marker beds; limestones with certain kind of fossils, quartz sandstones with trace fossils, hematite and bauxite beds, etc.
3. Key surfaces; unconformities, erosion surfaces on the bottom of channel sandstones, exposure surfaces of sandstone (with roots) or carbonate rocks (with fresh water calcites on the top), condensed beds, storm wave erosional surfaces (Fig.3), volcanic ash beds, and transgression onlapped beds(Fig. 4), etc.

Results of Correlation

The depositional cross section and stratigraphic correlation results are showed in Fig. 5, and the lithotypes distribution of each depositional stage is displayed in Fig 6.

CHARACTERISTICS OF MESOSI ∽ 'ENCE 9 (C_2)

Orthosequence set C_2^1

This orthosequence set consists of Os^1, Os^2, Os^3, and is characterized by containing *Fusulina - Fusulinella Zone*, equivalent to the Westphalian B, C, D deposits, Benxi Formation. Its thickness is about 30 m, and is developed during 307-303 Ma[12].This orthosequense set overlays the underlying Ordovician carbonate rocks. There is a hematite and bauxite claystone bed covering the unconformity surfaces. This claystone is a diachronic bed in the Paleozoic North China Basin, developing during the marine transgression. In northern Hedong Coalfield the claystone is overlaid by Os^1 mudstone with volcanic ash, but in the middle and southern Hedong Coalfield it is overlaid by Os^2 and Os^3 directly. O_s^2 and Os^3 are formed during the basin is flooding and deepening, for two lenticular beds of bioclastic limestones distributing on the top of each ortho-sequence, containing many marine fossils such as *Fusulina, Fusulinella*, crinoids, brachiopods, chaetetes and single carols, as well as horizontal burrows, represent the deposits of subtidal high energy shoals of open marine environments. The worm crawling traces and cracks on the top of the Os^3 implicate that it has been exposed before the C_2^2 Orthosequence Set deposits.

Orthosequence Set C_2^2

This orthosequence set includes 4 orthosequences,Os^4-Os^7, developing during 303-290 Ma[12] and is characterized by *Triticites and Montiparous Zone*. It is equivalent to Stephanian of Upper Carboniferous, Lower Member of Taiyuan Formation, with a thickness of 25-50 m. The Os^4 and Os^5 in northern are fluvial and deltaic deposits, with   erosional surfaces at the bottom of the channel sandstones, but to the middle and southern there are barrier and lagoon depositional systems distributed. The O_s^6 and Os^7 in northern are restricted marine deposits with semi-brackish water fossils in the lenticular limestone beds, in the middle area there are lagoon fine deposits, and tidal channel sandstones in the south area.

Considered in the 2nd order sequence of Mesosequence 9, the C_2^1 Orthosequence Set is transgressive system tract (TST), and the C_2^2 Orthosequence Set is highstand system tract (HST), the maximum flooding stage is the beginning of C_2^2.

CHARACTERISTICS OF MSOSEQUENCE 8 (P_1)

Include two sub-mesosequences: Mesosequence 8-1 and Mesosequence 8-2. The Ms 8-1 consists of Orthosequence Sets P_1^1 and P_1^2; and the Ms8-2 includes P_1^3 and P_1^4.

Fig. 3. Storm wave erosional surface on limestone L_2, the fresh water calcite indicates that the sea level of the shallow marine is close to the exposure environment (Sanjiao District, $P_1{}^1$).

Fig. 4. The marine silty-mudstone is onlapped on the tidal inlet sandstones, shows the marine flooding surface of the subsequence boundary (in Os^4 of $C_2{}^2$, Sanjiao District).

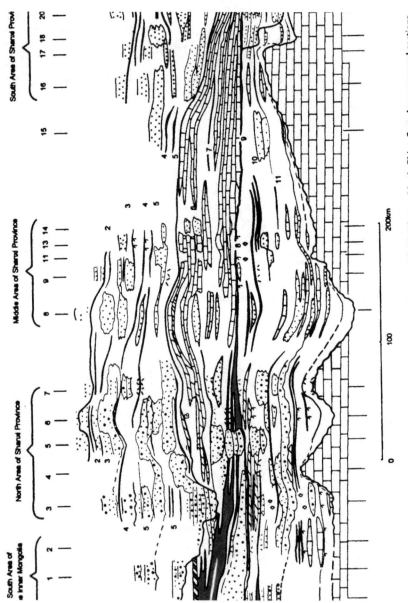

cross section of Permo - Carboniferous coal - bearing strata in the Hedong Coalfield, Western North China (based on measured sections

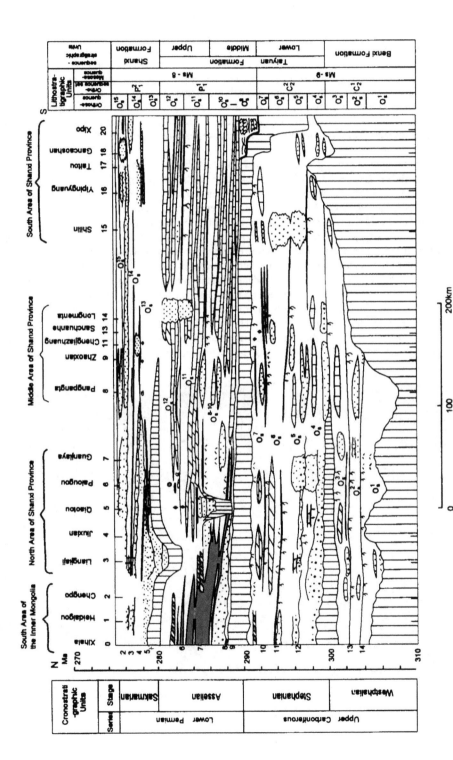

Figure 6 Depositional cross section of sequence stratigraphy in isochronic time scale shows the lithofacies distributions in the Hedong Coalfield,western North China.

Orthosequence Set P_1^1

This unit is the first orthosequence set of Early Permian, including the Middle and Upper Member of Taiyuan Formation. The geological time interval is 290-280 Ma, belongs to Asselian Stage. There are five orthosequences in P_1^1, including Os^8-Os^{12}. The Os^8 is an incised valley-fill, developing in southern Hedong Coalfield. The depth of the incised valley is more than 25 m. The filling-deposits in the valley are two cycles of tidal-fluvial channel sandstones and tidal flats of fine grained sandstones, siltstones and mudstones interbedded, with a coalbed on the top. An erosional surfaces on the bottom of the channel sandstones overlays the hematite and bauxite claystone which is formed from the paleosol of the underlying Ordovician carbonate rocks. In this section most of the Carboniferous deposits are eroded. It probably represents the incised valley-fill of the Early Permian Paleo-continental margin of the western North China Coal Basin.

The Os^9-Os^{12} mainly consist of carbonate rocks and interbedded with mudstones and coals, and their totally thickness is less than 40 m. The bioclastic carbonate rocks of Permo - Carboniferous in North China are formed in warm and shallow continental sea, no reefs have been found, and are developed as carbonate ramp. There are some differences in vertical sequences of the bioclastic carbonates between Os^9-Os^{10} and Os^{11}-Os^{12} in the middle portion of the research area, that is the former have coarsening upward sequences and the later have finning upward sequences. The carbonates in Os^9-Os^{10} are developed in lower energy deep water first and then in higher energy shallow water, the depositional accommodation becomes smaller along with the carbonate building up, indicates that the carbonate producing rate is quicker than the relative sea level rising rate. This probably is equivalent to the keep up style [10], implicating that in this case the relative sea level rises gradually which is benefit for coal developing. Coal seams are occurring along the shorelines while the relative sea level rises slowly. Whereas the carbonates in fining upward sequence of Os^{11} and Os^{12} are developed in higher energy shallow water originally, and than in lower energy deep water.   This indicates that the relative sea level rising rate is quicker than the

Fig. 7. The incised valley is filled with channel sandstones (Os^8), and a coalbed and limestone (Os^9) onlaps the sandstones. P_1^1, Xiangning county, southern Shanxi Province.

carbonate building rate, and the depositional accommodation becomes larger. This kind of carbonates possibly are equivalent to the catch up style [10]. In lateral all of these carbonates are pinching out northward along with the coalbeds of each orthosequence thickening and combining into a very thick coal seam with many kaolinite claystone interbeds in the northern Hedong Coalfield.

Orthosequence Set P_1^2

This is the second orthosequence set of Early Permian, including Os^{13}-Os^{15}, and is Shanxi Formation. Total thickness varies from 50 to 80 m. The time interval is from 280 to 275 Ma, belonging to Sakmarian. There is an incised valley occurring in northern Shanxi Province, with braided channel sandstones filling in (Section 3 on Fig. 2). This incised valley fill is different from the incised valley fill distributed in southern Shanxi Province, because there are double mud layers occurring in the foresets of cross bedding of the sandstones in the south, implicating the estuarine depositional environments developed in the paleo-continental margin, but the valley in the north distributes in front of the uplifting hills of the source area, and the filling sandstones are typical braided channel sandstones without any tidal influence criteria.

The depositional facies in the middle area are deltaic distributary channel sandstones and interdistributary bay filled with fine grained clastic deposits and coals. In the south there are quartz sandstones with coarsening upward sequence, formed in progradational clastic shorelines.

Considered in the second order sequence of the Mesosequence 8-1, the Orthosequence Set P_1^1 is transgressive system tract (TST), and the Orthosequence Set P_1^2 is highstand system tract (HST).

Orthosequence Set P_1^3

This is the lower orthosequence set of the Mesosequence 8-2, characterized by terrestrial gray green coloured clastic deposits, and is Lower Shihezi Formation, 50 -60 m thick, including Os^{16} and Os^{17}. Os^{16} consists of meandering channel sandstones with epsilon cross bedding and overbank fine deposits. Os^{17} consists of lacustrine and lacustrine-delta deposits. There is no coalbeds develops in this interval in the Hedong Coalfield.

Orthosequence Set P_1^4

This is the upper orthosequence set of the Mesosequence 8-2, and is Upper Shihezi Formation, characterized by gray green and dark red coloured rocks interbeds. According to the regional correlation with southern North China by sequence stratigraphy, it should be the most upper interval of Early Permian.

TRANSGRESSION AND REGRESSION

Two cycles of transgression and regression have influence in western North China, they are Westphalian transgression and Asselian transgression, and the second transgression is larger. The Westphalian (C_2^1) transgression comes from the east Pacific Ocean, through Northeastern China and North China, and finally to the west reaches the study area. The evidence of this transgression is that the earliest fossils of Upper Carboniferous, *Eostaffella subsolana* Zone occurs in Northeastern China, but not appears in North China. Then the shallow continental sea overlaps southward when the north Yinshan tectonic zone starts to uplift, which probably is caused by the southward subduction of the Mongolia-Siberian Plate. The volcanic ash beds increasing northward in the Hedong Coalfield should be related to this tectonic background. The wild spread of kaolinite paleosol existing on the top of Upper Carboniferous, shows that the sea level has dropped down and that the continental shelf is exposed after the southward regression.

The Asselian transgression is from the south to the north in western North China. The incised valley located in south Shanxi Province is filled during the relative sea level rising. From the cross section (Fig. 2),it can be found that the direction of the Permian transgression is from the south to the north, because there are only calcareous mudstones or marls and brachiopod in the north instead of marine limestones and fossils such as foraminifera, bryozoa, brachiopod in the middle and south area. These implicates the water is brackish in the north which is more close to the source area.

The limestones of Os^9-Os^{10} are thinning and pinching out to the north and also to the east of the middle North China Basin, and are not found in the southern North China Basin. It is suggested that the P_1^1 transgression occurs from the west Tethyan ocean first and passes through the incised valley in southern Shanxi Province, and then the east Pacific ocean transgression happens, from the southeast to the northwest.

COAL DEPOSITIONAL MODELS IN WESTERN NORTH CHINA

The coal seams occur underneath the limestones of Taiyuan Formation, so the depositional environment of the coal and that of the overlying carbonate are very close. Through the study, it has been found that the coal depositional model in transgressive system tract (TST) is different from that in highstand system tract (HST). TST consists of retrogradational subsequence sets, and HST consists of progradational subsequence sets [12]. In the continental sea of cratonic basin, the depositional rate is quite slow, the depositional processes are not continuous, and subsequences (4th order sequence) usually are very thin, sometimes it is difficult to divide and correlate in lateral, so we pay more attention to studying the distribution styles of orthosequences (3rd order sequence) or orthosequence sets (2nd order sequences).

Coalbed thickness, texture, distribution, and coal maceral composition are all controlled by the relative sea level changes. For example, in Mesosequence 8-1, coalbeds developed in TST (Orthosequence Set P_1^1) occur on the bottom of each orthosequence, and are overlaid by marine mudstone (several cm thick only) and limestone, and underlain by claystone. Coal is thicker in Os^9 and Os^{10} than in Os^{11} and Os^{12}, because the sea level rises slowly in the former and quickly in the later. From the basin (south) to the land (north), the coalbed thickness increases because the coals of each orthosequence combine together into a very thick coalbed (around 10 m) while the marine limestones become thin and pinch out . The main coalbed is on the bottom of the retrogradational orthosequence set (TST of P_1^1) of the Taiyuan Formation, and its texture is simple in the southern and middle Hedong Coalfield, no claystones intercalate in it. But to the north in front of the uplifting Yinshan Tectonic Zone, the combined thick coal intercalates with some kaolinite claystones, part of which are alternated from the volcanic ash. Coal thickness variations are mainly controlled by the relative sea level eustasy, but not related to the underlying channel sandstone distribution, because there is a mesosequence boundary between the coalbed and the sandstone, which indicates a depositional hiatus .

Coalbed of Shanxi Formation are formed in progradational orthosequence set of HST. Coal depositional environments are on the delta plain. Coal seam occurs on the top of coarsening upward sequence in each orthosequence. The upper coal seam is thicker and more pure, the lower coal seam with some claystones intercalated. That is because the later are developed on the lower delta plain whereas the former are developed on the transitional zone of the lower and upper delta plain, for the natural levee is low in the lower delta plain, more fine deposits are easy to be carried into the basin by crevasse [6]. Coal thickness variation are controlled by the underlying channel sandstones. Coalbed is thinning and pinching out to where the underlying channel sandstones are distributed. Coals are well developed in deltaic environments, along with the progradation the better coalbed horizon becomes higher to the south.

There are some differences of coal quality in lateral distribution and in different depositional system tracts. Bright and semi-bright coals increase to the south (basinward), whereas semi-dull and dull coals, as well as claystone interbeds increase to the north (landward). Coal maceral composition has the same variation trend. The bright and semi-bright coals are close to 90 % in the south, and 68.3 % averagely in the middle area, but only 33.5% in the north of the study area. The vitrinite content is about 85% in the south, and 68.5% (average) in the middle area, much less in the north. In Sanjiao-Liulin, the middle research area, 114 samples are picked from 11 coal pits. The simplified examined results are listed in the Table 1:

Table 1: Coal maceral compositions of Sanjiao-Liulin

	Vitrinite		Inertinite		Exinite	Mineral	Samples
	Vt	Vd	It	Im			
Coalbed 4	61.1	41.5	21.9	9.5	0.6	9.8	41
Coalbed 5	63.0	51.7	21.4	10.6	1.2	14.5	45
Coalbed 8	81.9	47.4	17.2	2.0	0.2	4.0	28

Vt - total vitrinite   Vd - desmocollinite   It - total inertinite   Im - macrinite

In the middle research area the Coalbed 8 is formed on the bottom of Orthosequence Set P_1^1 (TST), and the other two coalbeds are developed on the delta plain of Orthosequence Set P_1^2 (HST). Coalbeds 4,5,8 are formed in different depositional system tract of the 2nd order sequences. The Coalbed 8 of Taiyuan Formation in TST has the highest vitrinite content and lowest Inertinite and mineral content, possibly this is because that the coalbed is developed during the sea level rising and that the transgression occurs gradually, caused the marshes are always covered by water. The Coalbed 5 of Shanxi Formation in HST is developed on lower delta plain of subaqueous environments, that explains why it has higher vitrinite content and highest mineral and exinite content. Coalbed 4 is formed in the transitional lower delta plain, although its inertinite content is a little higher, vitrinite is still dominated and mineral matter content is low, coal quality is very good.

Coals produced from Sanjiao-Liulin area are well known of their hydrocarbon potential, this is the first successful place for coal methane extraction in China. The methane content of Coalbed 8 reaches to 14 - 20 m^3 .That is not only because the coals with low ash and high vitrinite content cause the cleats very well developed, but also because the desmocollinite is dominant in vitrinites. Coal petrologists consider that it is formed from the elementary plants developing in deeper water, usually with hydrogen-rich vitrinites. Macrinite is relatively higher in Coalbed 4 and Coalbed 5, which indicates that it has been through gelification in the beginning, and then oxidized when the water level of the peat swamps is dropped down by sediments progradation. That probably is one of the characteristics in the coals developed in HST.

SEQUENCE STRATIGRAPHY ANALYSIS IN TULUFAN-HAMI BASIN

Tulufan-Hami Mesozoic and Cenozoic Basin is located in the southeast of Urumqi, Xinjiang Uygur Autonomous. It is a foreland basin of the Tianshan tectonic uplifting zone, lies on the south foothills of Bogeda Mountain. The Lower and Middle Jurassic coal-bearing strata containing very thick coalbeds, and in recent years the petroleum geologists found that the exploited oil is originally generated from the coal [7].

Sequence Stratigraphic Unit Division
The outcrop investigations have done in Taibei depression,   central   Tulufan - Hami basin. Five

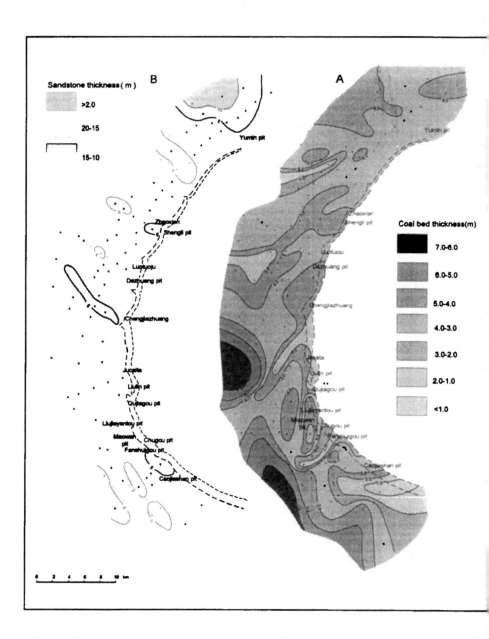

Figure 8. Isopach map of Coalbed 8 of P_1^1(A) and the underlying sandstone of C_3^2(B), shows the coal thickness variation is not controlled by the underlying sandstone distribution. Sanjiao-Liul8in area, middle Hedong Coalfield.

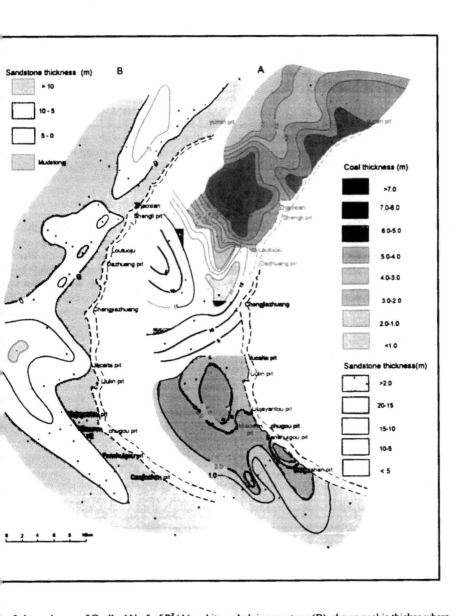

re 9. Isopach map of Coalbed No.5 of P_1^2 (A) and its underlying sanstone (B), shows coal is thicker where nder-lain sandstone is thin. In the middle of (A), the thickness of a post- channel sandstone is displayed, e the coals are eroded.Sanjiao Liulin, Hedong Coalfield.

measured sections (1:500) are showed on Fig. 10, the Sections 1,2,3 are along the north outcrop
of the basin and the Section 4,5 are along the Central Tectonic Uplifting Zone which is distributed
in the south of the Taibei depression. Total thickness of Jurassic coal measures is around 1000 m
which are consist of terrestrial clastic deposits. The measured sections including depositional
facies, sequence stratigraphy division, and depositional system tracts are all displayed in Fig. 10.
The Lower Jurassic (J_1) Includes Badaowan Formation and Sangonghe Formation. Badaowan
Formation mainly consists of alluvial fan coarse grained clastic deposits, workable coal seams and
lacustrine fine deposits are on the top of it. Sangonghe Formation consists of alluvial fan coarse
grained deposits on the lower part, and lacustrine fine deposits on the upper part, and coalbed is
not common.

The Middle Jurassic (J_2) includes Xishanyao, Sanjianfang, and Qiketai Formations. Coal seams
occur in the middle of Xishanyao Formation, and the other two formations consist mainly of
lacustrine fine grained clastic deposits and mudstones, and with some calcareous deposits
occurring on the top. The early Upper Jurassic Qigu Formation is overlain the Mesozoic coal-
bearing strata. The Qigu Formation is mainly consist of lacustrine dark red mudstones which can
be as thick as more than 500 m, and is an very good and important cover bed for the petroleum
generating from the coals and preserved till now.

Characteristics of sequence stratigraphy
There are unconformities represented by the erossional surfaces and coarse grained clastic deposits
lying on the sequence boundaries. Each formation consists of an orthosequence set, which
includes one or several orthosequences. Commonly there are fining upward sequence deposits
occurring in the lower part of the orthosequences, and coarsening upward sequence deposits in the
upper part. The orthosequence are consisted by subsequences. The fining upward sequence
represents the retrograda-tional subsequence set, which is developed while the lake level of this
basin rising and lacustrine transgression occurring. The enlarged lake environment forced the
alluvial fan retrogradation and which is covered by lacustrine water and becomes into subaqueous
fan. Then the retrogressive sequence is developed, which is corresponded to the TST of paralic
strata. The coarsening upward sequence in the upper part of an orthosequence represents the
progradation process of alluvial fan or braided channel which are flowed into the large lake.

Compare with the paralic coal-bearing strata in western North China, coalbeds occur on the upper
part of the retrogradational subsequence set (TST), but not on the bottom of orthosequence set.
That is because the thickness of subsequence is too thin in the cratonic basin, so the basic research
unit there is orthosequence but not subsequence. The other reason is the shallow water of marine
or lacustrine shoreline environments are the best place for marshes and peat swamps developing
and coal-forming. In brief, the most important condition for thicker and better coals forming is the
ground water surface gradually and continuously rising and the swamps permanently filling with
shallow water.

According to the correlation of the out crop measured sections and the drilling geophysical logs,
the debris flow of alluvial fan and braided channel fill deposits are commonly occurred in the
north of Section 1,2,3, and the paleo-current direction are mainly from the north to the south, so
the source area of the sediments should be in the north of the basin. In Section 4,5, where the
Central Uplifting Tectonic Zone is located now, there are more lacustrine fine grained deposits,
implicates that there are the originally deeper water area of the ancient lake during the Jurassic
coal-forming time.

Characteristics of the Jurassic Coals
Coals of Lower Jurassic Badaowan Formation are mainly developed in the west of Tulufan, but
not important in the research area of Taibei depression. Thick coalbeds occur in Middle Jurassic

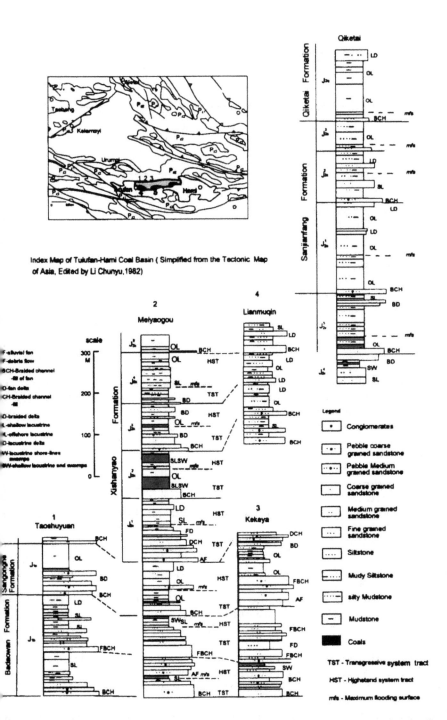

Figure 10. Measured outcrop sections of the Lower and Middle Jurassic coal-bearing strata in Tulufan-Hami Basin, lithofacies types, depositional system tracts and the sequence stratigraphic units are indicated on the sections.

Xishanyao Formation (J_2x), which can be divided into five orthosequences, and the thicker coalbeds occur in the 2nd and 3rd Orthosequences. Total thickness of the coalbeds in J_2x ranges from 50-70 m, and the kaolinite claystone interbedded increased northward. The coal maceral composition and the index of coal facies are listed in Table 2.

Table 2: Coal maceral composition and coal facies index of Tulufan-Hami Basin

	Vitrinite		Inertinite		Exinite	GI	TPI
	Vt	Vd	It	Im			
Meiyaogou(J_2x)	72.8	13.0	23.0	4.6	4.2	3.38	4.26
Qiketai (J_2x)	65.5	41.0	26.3	4.0	8.2	3.23	0.66
Kekeya (J_1b)	75.1	17.0	22.5	2.9	2.4	4.15	2.83

Vt - total vitrinite   Vd - desmocollinite   It - total inertinite   Im - macrinite

The index of coal facies analysis are presented by C.F.K.Diessel [4]:
GI is gelification index, and TPI is tissue preserved index.

$$GI = \frac{Vitrinite + Macrinite}{Fusinite + Semi\text{-}fusinite + Inertonertinite}$$

$$TPI = \frac{Telinite + Telocollinte + Semifusinite + Fusinite}{Densinite + Macrinite + Inertonertinite}$$

From the lithofacies distribution of the Middle Jurassic, the J_2x coals produced from Qiketai, represent the deep water coal-forming environments, and the coals from Meiyaogou represent shallow water environments. The coal petrology research results support this assume. Compared with the coals from Meiyaogou, the Qiketai coals are 3 times rich in desmocollinite content, exinite and inertonertinite contents are twice, macrinite content is 5 times, and the TPI index is much lower. All the desmocollinite, exinite, inertonertinite and macrinite etc. are commonly distributed in deeper water. That is explained why the TPI index is low.

CONCLUSIONS

1. The depositional rate is quite slow in the North China cratonic basin, from 310 to 275 Ma, the Paleozoic coal measures are only about 140m thick, but coalbeds can be as thick as 10 m. In Tulufan-Hami foreland basin the J_1-J_2 coal measures are ranged from 1000 to 1300 m during about 45 Ma, the depositional rate is much quicker.
2. In western North China two orthosequence sets including 7 orthosequences are recognized in Upper Carboniferous, and 4 orthosequence sets including 13 orthosequences are divided in Early Permian. In Tulufan-Hami basin there are 3 orthosequences in Early Jurassic, and 9 orthosequences in Middle Jurassic.
3. There are two transgression and regression cycles occuring in C_2-P_1 in the North China basin. The Westphalian transgression is from the east (Pacific ocean) to the west, and the Asselian transgression is from the south (Tethys from the southwest first and then Pacific from the southeast) to the north.
4. Two depositional system tracts are common in these two paralic and terrestrial basins, they are TST and HST, and we consider the incised valley fill as the bottom deposits of TST [12] in the beginning of $P_1{}^1$. TST is represented by retrogradational sequence sets (Os or Ss), and HST is represented by progradational sequence sets.
5. Coals in these two basins have high hydrocarbon potential (methane in western North China and petroleum in Tulufan-Hami), and the reason is that they are rich in vitrinites, especially in

desmocollinites, which have weak fluorescence and are formed in deeper water, and that the mineral matter content is low. These kind of coals are mainly developed in TST, because during transgression or sea/lake level rising, the altitude between the base level of the basin and the source area decreases, causing the sediments supply decrease and so vitrinite-rich coals are formed.

6. Coalbeds formed in TST may be used as a new coal depositional model, the predictive coal accumulation regularities are that the coal thickness increases from the basin to the land, and is not controlled by the underlying channel sandstone distribution, and that hydrocarbon-rich bright coals decrease landward.

ACKNOWLEDGMENTS

Prof. Pan Zhiguei, Chu Zhimin and graduate students Zhang Shengli, Fang Dequan, Zhao Handi, Ye Hebin, Ma Shimin, Zhang Huiliang, Zhang Yundong participate part of the works. We are grateful to them. We appreciate Academician Professors Wang Hongzhen and Yang Qi for their directs and help. Many thanks to Prof. Chen Zhonghui, Shang Guanxiong, Kong Xianzhen, Wang Changgui, and Liang Shijun who have given us many helps. We also thank the Shanxi Coal Geology and Exploration Bureau, North China Petroleum Geological Bureau, Shanxi Geological and Mineral Resources Bureau and the Tulufan-Hami Petroleum Geological and Exploration Campaign Headquarters for giving us the opportunities to complete these research works.

REFERENCES

1. Burchette,T.P. and V.P.Wright, Carbonate ramp depositional systems, In: Sedimentary Geology 79, 3-57, Elsevier Science Publishers (1992).
2. Chen Zhonghui et al., Depositional Environments and Coal-Accumulating Regularities of the Paleozoic Coal-Bearing Strata in East Ordos, China University of Geosciences Publishing House (1989), in Chinese.
3. Chen Zhonghui et al., Depositional Environments and Coal - Accumulating Regularities of the Paleozoic Coal-Bearing Strata of North China, China University of Geosciences Publishing House, (1883), in Chinese.
4. Diessel, C.F.K., *Coal-Bearing Depositional Systems*, (1992).
5. Ferm, J.C. and G.A. Weisenfluh, Evolution of some depositional models in Late Carboniferous rock of the Appalachian Coal Fields, *International Journal of Coal Geology*, pp. 259-292, (1989).
6. Horne, J.C., J.C. Ferm, and B.P. Baganz, Depositional models in coal exploration and mine planning in Appalachian region, *Bull. of AAPG*, Vol. 62, No. 12 (1978).
7. Huang Difan, Zhang Dajiang, Li Jinchao,Huang Xiaoming, Hydrocarbon generated from the Jurassic coal-bearing strata of Tulufan Basin, China, In: Oil Generated from Coal, the Advances of Geochemistry, Huang Difan et al., eds. Petroleum Industry Publishers (1992), in Chinese.
8. Li Yunlan, Biostratigraphy of the Carboniferous and Permian, In: Coal-Accumulating Regularities and Coal Resources Assessment in Hedong Coalfield, Gui Xue zhi ed. Shanxi Publishing House of Science and Technology, (1993), in Chinese.
9. Mitchum,R.M., P.R. Vail and S. Thompson, Seismic stratigraphy and globle changes of sea level, Part 2: The depositional sequence as a basic unit of stratigraphic analysis, In: *Seismic Stratigraphy - Applications to Hydrocarbon Exploration*, C.E. Payton (ed.) AAPG Memoir, 26, 53-62 (1977).
10. Sarg, J.F., Carbonate sequence stratigraphy, In: Sea-Level Change-An Integrated Approach: SEPM Special Pub. (1988).
11. Vail, P.R. et al., The stratigraphic signatures of tectonics, eustasy and sedimentology-- an overview, In: Cycles and Events in Stratigraphy, ed. by Gerhard Einsele, Werner Ricken and Adolf Sailacher (1991).
12. Van Wagoner, J.C., R.M. Mitchum, K.M. Campion and V.D. Rahmanian, *Siliciclastic Sequence Stratigraphy in Well logs, Cores, and outcrops: Concepts for High-Resolution Correlation of Time and Facies.* AAPG Methods in Exploration Series, 7. (1990).
13. Wang Zhenyin, Liu Hannan, Tang Jianxiu, Shang Guanxiong, Division of the Paleozoic coal-bearing strata in North China Platform, In: *Geology and Mineral Resources Records of North Chine*, Vol. 11, No.1 (1996), in Chinese.
14. Wang Hongzhen and Shi Xiaoying, A scheme of the hierarchy for sequence stratigraphy, In: *Sequence Stratigraphy and Sea-Level Changes of China, Earth Sciences*, Vol. 7, No. 1 (1996).

Proc. 30th Int'l Geol. Congr., Vol. 18, Part B, pp. 21-31
Yang Qi (Ed.)
© VSP 1997

Petrology and Depositional Environment of Early Jurassic Coal, Western Australia

KRISHNA K SAPPAL
School of Applied Geology, Curtin University of Technology, Perth, Western Australia, 6001
and
NANA SUWARNA
Pusat Penelitan dan Pengembangan Geologi, Bandung, Indonesia

Abstract

The Early Jurassic sub-bituminous coal of the Hill River area located in the Dandargan Trough of the northern Perth Basin Western Australia has been fully explored since 1981. A total coal resource of approximately 450 million tonnes has been delineated for open cut mining in the near future. The vitrinite reflectance ranges from 0.47% to 0.50% which places the coal in sub-bituminous rank as per Australian standards. The coal has a high ash and moisture content and it is considered suitable fuel on or near site power station. Similar to the Permian coals of Western Australia, the coal is finely laminated with a subdued lustre and the dominant lithotypes are dull banded, dull and banded with minor bright and bright banded. The dominant maceral groups of the coal are vitrinite and inertinite followed by mineral matter and the exinite group of macerals. The mineral matter associated with the macerals consists of clay, pyrite and quartz. The sulphur content in the coal from the borehole CPCH1 ranges between 0.92% to 2.36%. On the basis of petrographic analyses, type of pyrite and trace elements distribution in the coal samples of the bore hole CPCH1 of the Mintaja block, the depositional environment postulated for the coal is an upper to lower delta type with a regressive phase of a marine transgression. This model for the depositional environment is also supported by the association of acritarchs and palynomorphs in the coal measures of the Hill River area.

Key words : Early Jurassic, coal, macerals, minerals and depositional environment

INTRODUCTION

The Early Jurassic coal in the Hill River area, located 250 km north of Perth in the Dandargan Trough of the Perth Basin was first intersected in 1964 during the drilling for oil in the Perth Basin. The regional setting of the Hill River area is given in Fig.1. Since 1981, the CRA Exploration Pty Ltd (CRAE) has undertaken exploration and drilling in the area, and a total mineable resource of coal approximating 450 million tonnes has been proved. In 1990 the company submitted a tender to supply coal based power from the area to the State Energy Commission of Western Australia based on a power station of capacity of 600 megawatts.

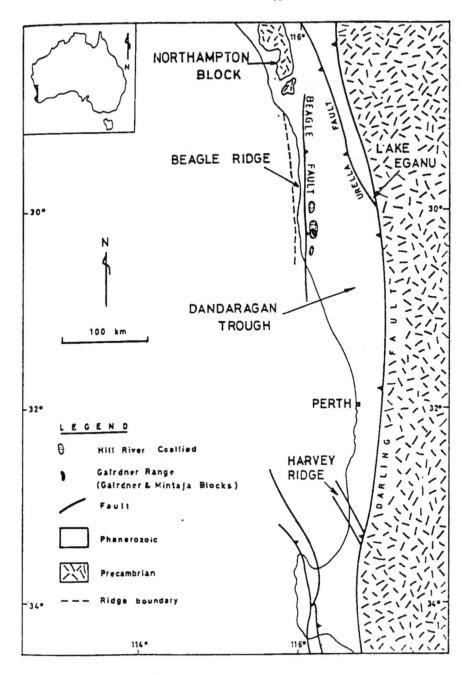

Figure 1. Regional geological setting and locality of the Hill River Coal-field, Western Australia

The coal samples for this study were taken from the five sub-seams G1 to G5 of bore hole CPCH1 drilled in the Mintaja block of the area. The particulate samples of the coal were prepared and examined according to the Standard Association of Australia Codes [1,2,3]. This study is a component of the research program being undertaken at the Coal Research Laboratory, Curtin University of Technology, Perth for petrographic and geochemical characterisation of Western Australian coals.

GEOLOGICAL SETTING

The coal samples for the study were obtained from the bore hole CPCH1 drilled by the CRA Exploration Pty Ltd (CRAE) in the Mintaja Block of the Gairdner Range, located in the Hill River coalfield (area), Northern Perth Basin, Western Australia, [Fig.1]. The Perth Basin is a deep linear trough extending along, north-south trend over a distance of 1000 km. The eastern boundary of the basin is marked by major tectonic feature named the Darling Fault with a continental slope to the West. According to Cockbain [4] the basin is polycyclic containing approximately 15,000 metres of Silurian to Quaternary sequences. The Silurian to Early Neocomian sequences consisting mostly of shallow marine and fluvial sediments which were deposited in an interior fracture setting. And the late Neocomian to Quaternary sequences again consisting of shallow marine and fluvial sediments were laid down in a marginal sag basin. The Dandargan Though is a half graben and respresents the largest subdivision of the Perth Basin. The through is filled with 15,000 metres of mostly Permain to Jurassic clastic sediments with minor association of coal.

The Hill River coalfield located in the Dandaragan Trough has an areal extent of approximately 100 sq km, and the economic coal seams are restricted to a relatively narrow stratigraphical interval of Early Jurassic age, described as the Cattamarra Coal Measures, Playford et al and Cockbain [5, 4]. The coal measures have also been described as the Hill River coal measures as per the nomenclature adopted by CRA Exploration Pty Ltd (CRAE), Kristensen [6]. The coal measures sub-crop and out crop over a north trending strip of 80 km in length and 5 to 10 km in width. They extend from Eneabba in the North to Wongonderrah in the South, in an area drained by the Hill River and the Cockleshell Gully, [Fig. 2]. The coal measures in the area are underlain by the fluvial Lesueur Sandstone of Late Triassic age, and conformably overlain by the fluvial Yarragadee Formation of Late Jurassic age. The coal measures are bounded by the Lesuer Fault to the west and Warradarge Fault to the east.

Extensive exploration and drilling in the Gairdner and Mintaja Blocks of the coalfield has delineated a total coal resources of 450 million tonnes which represents the largest early Jurassic coal deposit in Western Australia. The economic seam in the bore hole CPCH1 is named G and it has six sub seams G1 to G6 out of which G1 to G5 are mineable and reported in the paper. The coal seams occur about 20 metres below the surface in the northern edge of the coalfield and dip towards the south. The CPCH1 bore hole contains 34 metres of coal bearing sediments, with four intervals of coal with a cumulative thickness of 16 metres. The coal measures consists of cyclical sandstones, fining upward into interbedded mudstone and shale and coal. The sand-shale ratio gradually increases towards the base of the coal measures, where fine grained sandstone with some zones of marine and brackish water microplanktons and shell beds occur, Sapppal et al [7].

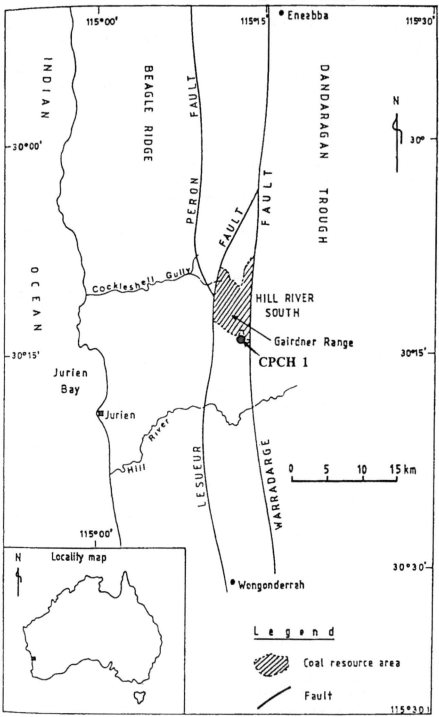

Figure 2. Tectonic setting and location of the bore CPCH1

PETROLOGY

Coal is a heterogenous combustible sedimentary rock, and much of the petrological investigations involve characterisation of heterogeneity. This heterogeneity of the coal is visible macroscopically as its banding and variation in brightness, however microscopic examination reveals the details of organic constituents (macerals and microlithotypes) and the discrete inorganic mineral matter. The recent development of standards in preparation and examination techniques of the particulate polished sections, and the international agreement in definitions of organic components of coal has given greater impetus and the heterogeneity can now be quantified. This has allowed the assessment of coal for processing suitable to utilization with lesser unfavourable impact on environment. The macroscopic (lithotypes) contents of the sub-seams G1 to G5 of the bore hole CPCH1 are given in Table 1 and the data on maceral groups, mineral matter, vitrinite reflectance and the total sulphur in the coal samples is given in Table 2. The data in both these tables represent averages of samples from a number of sub-sections of the individual sub-seams of the bore hole.

Lithotypes

The terminology used to describe northern hemisphere coals as vitrain, clarian, durain and fusain is not equally applicable to Gondwana Coals of southern hemisphere due to their finer laminations and subdued lustre and also because the Gondwana coals of southern hemisphere have been formed under a different set of geological, biological and geographical conditions than the coals of similar age in northern hemisphere. A smaller limit of a minimum of 3 mm thickness for laminae and their brightness is used to describe and classify the lithotypes. The terms applied to the lithotype analysis are dull, dull banded, banded, bright banded, bright and fusain for fibrous coal, as per definitions of Diessel [8]. The terms are based on thickness of laminae and their brightness. The bright component of dull coal is 0-10%, dull banded coal 10-40%, banded 40-60% bright banded coal 60-90% and the bright coal 90-100%. The term fusain for fibrous coal is used similar to the one used for northern hemisphere coals.

Table 1. Lithotypes content (linear %) of the sub-seams, CPCH1

Sub-seams Thickness (m)	Bright	Bright Banded	Banded	Dull Banded	Dull	Fusain	Clastics
G1/1.2	20	12	16	39	11	2	0
G2/4.4	6	8	22	35	29	0	0
G3/1.3	14	32	10	29	3	6	6
G4/1.1	3	8	13	25	21	30	0
G5/0.4	0	5	11	0	84	0	0

Table 2. Maceral, mineral matter, vitrinite reflectance and sulphur content of the sub-seams CPCH1

Sub-seam	Vitrinite %	Exinite %	Inertinite %	Minerals %	Ro_{max}%	Sulphur %
G1	61	13	22	4	0.47	0.92
G2	62	12	22	4	0.49	1.42
G3	63	14	16	7	0.50	1.17
G4	52	20	19	9	0.49	2.36
G5	55	16	15	13	0.50	N/A

The proportions of lithotypes vary from sub-seam to sub-seam as shown in Table 1, which includes the percentages of the aggregate thicknesses of each lithotype and clastic partings in an individual sub-seam, calculated on the basis of their proportions. The sub-seam G1 is dominated by dull banded type (39%), and the sub-seam G5 has no dull banded lithotype in its composition. Overall the sub-seams are dominated by dull banded, banded and dull lithotypes with the exception of the sub-seam G5 which has 84% of the dull lithotypes. The variations in the lithotype composition of sub-seams are associated with the frequency of changes in the water table during the peat formations, as is observed from the lithotype profiles of the individual sub-seams, Suwarna [9]. In sub-seam G1, the dull lithotypes occur in close association with clastic layers. In the individual sub-seams brighter lithotypes usually grade with duller lithotypes. This upward grading into brighter ones and clyclicity in the individual sub-seams profiles indicates changes in the water table during the deposition of peat as a consequence of varying rates of subsidence in the mire area. The bright and bright banded lithotypes are formed due to flooding in the mire at a relatively shallow depth. The dull lithotype is formed under somewhat deeper water, whilst fusain is formed in drier conditions combined with forest fires. The fusain bands in coal represent dry conditions followed by wet conditions for bright, bright banded, banded and dull coals and the wettest conditions for carbonaceous shale and clastic layers in the coal seams.

Macerals

The macerals in the coal are products of peat forming flora, depositional environment and coalification in the sedimentary basin. Based on the physical and chemical properties, macerals are described under three groups, namely vitrinite, exinite and inertinite. Each of the maceral group is sub-divided into a number of macerals based on optical properties. The maceral composition of the coal is given as maceral groups and mineral matter in Table 2. The proportions of maceral groups and minerals are calculated from the sections of the individual sub-seams and these are given as percentages in the table. The Table also contains data on vitrinite reflectance and the total sulphur in the coal. The maceral composition of the sub-seams based on the mineral matter free basis is also presented in Fig.5.

The vitrinite group occurs as microbands, disintegrated and fragmentary in the coal. The macerals of the group recorded in the coal are telinite, telocollinite, desmocollinite, corpocollinite and vitrodetrinite. Fig. 3 illustrates telocollinite with cleats and desmocollinite associated with sporinite, cutinite, semi fusinite and inertodetrinite. As shown in Table 2, the vitrinite content of the five sub-seams ranges from 52% to 63% and the vitrinite reflectance has a range of 0.47% to 0.50%. According to the vitrinite reflectance the coal is classified as sub-bituminous as per Australian Standards.

The exinite group of macerals recognised in the coal consist of sporinite, cutinite, alginite, liptodetrinite and resinite. The alginite bodies in the coal are shown in Fig. 4 along with sporinite and resinite. The sporinite and cutinite association with desmocollinite is shown in Fig. 3. The content of exinite group in the coal ranges from 12% to 20%, which is higher than the one for Permian coal of Western Australia, Sappal et al [10]. The higher exinite content is postulated due to marine influence in depositional environmental of the coal. This environment has also imparted higher sulphur (Table 2) content to the coal due to the presence of framboidal pyrite, Fig. 4.

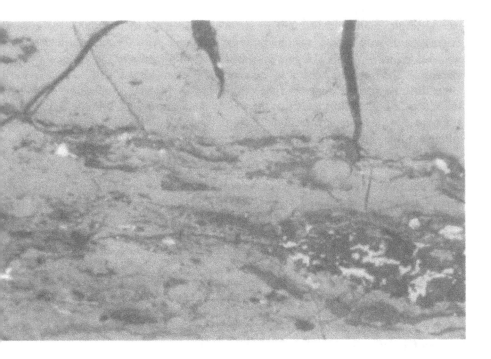

Figure 3. Vitrinite (grey) association with inertinite (white) and exinite (dark grey), oil immersion x500

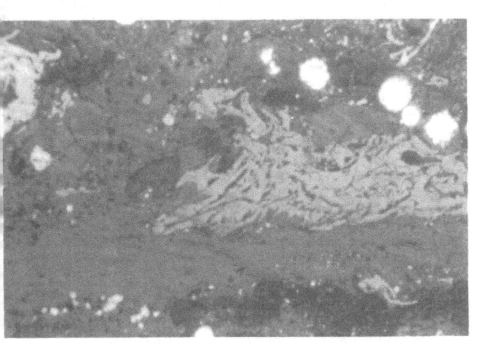

Figure 4. Rounded framboidal pyrite (white) association with exinite (dark grey), vitrinite

The maceral group inertinite in the coal is mainly dominated by semifusinite inertodetrinite and fusinite with minor micrinite, sclerotinite and macrinite. Fig.4 shows association of semifusinite, micrinite and inertodetrinite with alginite and telocollinite Rounded grains of framboidal pyrite are associated with inertinite and vitrinite group of macerals. The inertinite content in the coal ranges from 15% to 22%, which is much lower than the vitrinite content in the coal, Table 2. Maceral composition of the five sub-seams in Fig.5   illustrates that the coal is dominated by the vitrinite and inertinite groups, due to the predominance of dull banded, banded and dull lithotypes which contain vitrinite as fine and disintegral entities.

Minerals

In addition to the organic components, the coal also contains inorganic constituents as the mineral matter. The quantity of inherent minerals in the coal are not reported here, however, adventitious minerals as discretes have been identified during the microscopic investigations. The clay minerals occur in association with the macerals (Fig. 3) and pyrite minerals occur as framboidal rounded grains (Fig. 4) in the coal.

Minor amounts of quartz and carbonates are also present in the coal. The discrete minerals present in the coal range between 4% to 13% as shown in the Table 2. The presence of framboidal pyrite and veinlets of pyrite in the coal accounts for higher sulphur content with a range between 0.92% to 2.36%, which is classified as medium to high category for Australian coal, Hunt and Hobday [11].

The framboidal pyrite and veinlets of pyrite in the coal are formed during the depositional and diagenetic phases of the precursor peat formation, Weisse and Fyfe [12]. The framboidal pyrite (Fig.4) is present as clusters or individuals upto 220 microns in size, and this pyrite along with pyrite cell fillings represents syngenetic pyrite. The isolated pyrite in cleats and fracture infillings in the coal is of epigenetic origin.

Trace Elements

Besides discrete mineral associations with the macerals the coal also contains a large number of trace elements which form organic, inorganic and mixed organic-inorganic compounds, these are related to depositonal setting and parent plant material, Diessel [13]. The concentration of trace elements in the coal varies from less 1 ppm to 3140 ppm. The trace elements B, Cr, Mn, Cu, Sr, Zr, Mo, Cd and Th predominantly occur as organic affinities, while the Be, V, Co, Zn, Ga, Y, Pb and U are present as inorganic affinities. The Ni and As are interpreted to be present as both organic and inorganic associations, Suwarna [9]. Swaine [14] considered that higher concentrations of boron (B) in coal as an indicator of marine influence in depositional environment. The concentration of boron in the Hill River coal ranges between 37 ppm to 409 ppm, and sub-seam G3 of drill hole CPCH1 contains the highest concentration which suggests presence of marine influence during the deposition of the Hill River Coal Measures in the Dandargan Trough of the Perth Basin.

From the view point of impact on environment, certain trace elements, such as As, Cd, Cr, Ni and Pb produce potential hazards and these have to be taken into consideration because of their affect on humans, animals and plants Gehr *et al* [15].

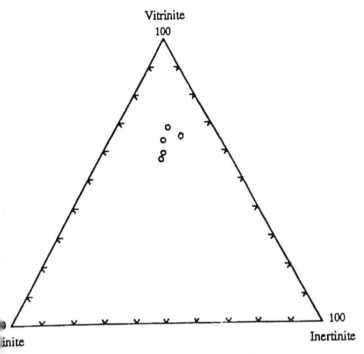

Figure 5. Maceral composition of the coal (mmf)

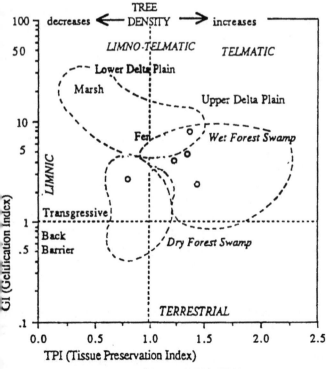

Figure 6. Facies diagram of the coal (CPCH1)

DEPOSITIONAL ENVIRONMENT

The interpretation of the depositional environment of the coal from the bore hol
CPCH1, Mintaja Block, Hill River coalfield is based on maceral analyses and th
petrographic indices defined by Diessel [16, 13]. The two petrographic indices as show
in the Fig.6 are Gelification (Gl) and the Tissue Preservation Indices (TPI). The Gl is
measure of the persistence of wet conditions in the environment and it is a ratio ɑ
vitrinite plus macrinite, and semi fusinite plus fusinite and inertodetrinite. An increase i
Gl indicates wetter conditions. The TPI is a measure of the predominance of macera
with remant cellular structure over the ones without cellular structure, and thus expresse
as a ratio between telinite, telocollinite, semi fusinite plus fusinite over desmocollinitɛ
corpocollinite, inertodetrinite plus macrinite. Higher TPI demonstrates decomposition ɑ
peat by oxidation in terrestrial forests. The plot of the Gl and TPI for the coal is give
as the facies diagram in Fig. 6. The diagram shows that the coal was deposited in wɛ
forest swamp in the upper delta plain and marsh of the lower delta plain during
regressive phase of a marine transgression during the Early Jurassic. The presence ɛ
marine influence in the depositional environment is also supported by the highɛ
concentration of boron, abundance of framboidal pyrite and acanthomorph acritarchs in th
coal measures, Sappal et al [7].

CONCLUSIONS

On the basis of lithotype and maceral analyses, and the distribution of mineral matte
particularly association of framboidal pyrite, the depositional environment for the coɑ
from the bore hole CPCH 1, Perth Basin, Western Australia is interpreted as an upper tɛ
lower delta type with a regressive phase of a marine transgression. The model is alsɛ
supported by the sedimentary structures present in the coal measures and the associatioɪ
of acanthomorph acritarchs with pollen and spores Sappal et al [7]. The petrologica
investigation of the coal has an impact on the exploration and utilization of coal for th
power generation.

REFERENCES

1. Standards Association of Australia. Code of Practice for preparation of hard coa samples for microscopic examination by reflected light. *AS 2061.* (1977).
2. Standards Association of Australia. Determination of the maceral group compositioɪ of bituminous coal and anthracite (hard coal). *AS 2515.* (1981).
3. Standards Association of Australia. Terms relating to the petrographic analysis oɪ bituminous coal and anthracite (hard coal). *AS 2418(5).* (1982).
4. A.E. Cockbain. Perth Basin, in Geology and Mineral Resources of Westerɪ Australia. *Western Australia Geological Survey, Memoir 3*, pp. 495-524. (1982).
5. P.E. Playford, A.E. Cockbain and G.H. Low. Geology of the Perth Basin, Westerɪ Australia. *Western Australia Geological Survey, Bulltetin* **124**. (1976).
6. S.E. Kristensen. *Hill River Coal Project Regional Geology.* Unpublished CRAE report. (1989).
7. K.K. Sappal. and A. Islam. Petrology and palynology of Cattamarra coal, Pertl Basin, Western Australia. *Geological Society of Australia, Abstracts*, **32**, p.132 (1992).
8. C.F.K. Diessel. Correlation of macro and micropetrography of some New Soutl Wales coals. *Proceedings of the Eighth Commonwealth Mining and Metallurgica Congress*, 6, pp. 669-677. (1965).

9. N. Suwarna. Petrology of Jurassic Coal, Hill River Area, Perth Basin, Western Australia. *PhD Thesis (unpublished), Curtin University of Technology.* (1993).

10. K.K. Sappal and B. Santoso. Petrology of selected Early Permian coal from Collie and Perth Basins, Western Australia *Proceedings Thirtieth Newcastle Symposium,*.221-228. (1996).

11. J.W. Hunt and D.K. Hobday. Petrographic composition and sulphur content of coals associated with alluvial fans in the Permian Sydney and Gunnedah Basins, Eastern Australia, in Sedimentology of Coal and Coal bearing Sequences. *International Association of Sedimentologists, Special publication* 7, pp.43-60. (1984).

12. R.G. Wiesse and W.S. Fyfe. Occurences of iron sulphides in Ohio coals. *International Journal of Coal Geology,* 6 pp 251-276. (1986).

13. C.F.K. Diessel. *Coal-bearing Depositional Systems.* Springer-Verlag. (1992).

14. D.J. Swaine. Boron in coals of the Bowen Basin as an environmental indicator. *Proceedings Second Bowen Basin Symposium, Geological Survey Queensland* Report, **62**, pp 41-48. (1971).

15. C.W. Gehr, D.S. Shriver, S.E. Herbes, E.J. Salmon and H. Perry. Environmental health and safety implications of increased coal utilization, in Elliot, M.A. (Ed) *Chemistry of Coal utilisation,* New York Willey, pp 2159-2233. (1981).

16. C.F.K. Diessel. An appraisal of coal facies based on maceral characteristics. *Australian Coal Geology,* **4**, pp.474-484. (1982).

Proc. 30th Int'l Geol. Congr., Vol. 18, Part B, pp. 33-46
Yang Qi (Ed.)
© VSP 1997

Depositional Evolution and Coal Accumulation of Ordos Basin

WANG SHUANGMING , LU DAOSHENG, ZHANG YUPING

Shaanxi Coalfield Geology Bureau, Xi' an, 710054, Shaanxi Province, China

Abstract

Ordos Basin, a huge intracontinental depression, influenced by Pacific structural domain together with Tethys, is formed during Mesozoic Era. Evolutional process of the depression can be divided into four epoches: Early Jurassic to early period of Middle Jurassic, Middle Jurassic, Late Jurassic and Early Cretaceous, of which, Early Jurassic to early period of Middle Jurassic is an important coal – accumulating period. Coal – accumulating area, as a ring, is distributed round subsidence centre of the basin, and in which numbers, thickness and change regularity of seam vary with different position of the basin. It is showed by tectonics and depositional analysis that the east boundary of the basin is situated to the east of Datong to Yima, and that perfect match of tectonic turning period with palaeoclimate in favour of formation for plant remains to a great extent is an essential factor controlling formation of the seam.

key words: basin, coal – accumulating period, coal – accumulating area, tectonic turning period

INTRODUCTION

The research project covers the Ordos Basin and its periphery region with an area of about 400 ,000 Km². The basin contains three coal –forming sequences of Carboniferous – Permian, Triassic and Jurassic, in which Jurassic is the most significant. The general coal reserve reaches 2000 billion tons. Ordos Basin is in the forefront of the world – grade bearing basins.

With well –developed and huge thickness of sedimentary strata, Ordos Basin can be grouped into three tectonic –depositional evolutional stages: Early Palaeozoic, Late Palaeozoic to Triassic and Early –Middle Jurassic to Early Cretaceous. Different stages are formed in certain geotectonic settings respectively. It is concluded that coal – accumulating process only takes place during Late Palaeozoic, Late Triassic and Early to Middle Jurassic, of which the coal – accumulating process during Early to Middle Jurassic focuses one's attention upon.

Early Palaeozoic, a development stage of shallow – sea basin within intraplate
During Early Palaeozoic, Ordos area(meaning nowadays distribution scope of Ordos Basin)belongs to a part of continental plate of North China, and on the both sides of south and north is confined by Qinling – Qilian and Xingmeng trench respectively. An

aulacogen is developed on the west side. This geological period is characterized by mutual opposites and common development between shallow – sea basin of stable intraplate and movement belt of continental margin[1] . During late period of Early Palaeozoic, subduction of oceanic crust occurs successively on south and north sides of continental plate of North China, and Caledonian folded belt is formed along the continental margin, This compressional process of subduction of oceanic crust resulted in, as a whole, uplift of continental plate of North China, and Caledonian folded belt is formed along the continental margin, This compressional process of subduction of oceanic crust resulted in, as a whole, uplift of continental plate of North China and consumption of shollow – sea basin of intraplate. Then, the area is undergone uplift and denude for long.

Late Palaeozoic to Triassic, a development stage of huge intracontinental basin
The connection of shollow – sea basin within intraplate of North China during Late Palaeozoic with outplate ocean basin is separated by the Caledonian folded belt, as a folded orogenic terrain of continental margin of North China. Thus, a depositional basin, represented by the epicontinental sea, is formed, which takes Caledonian folded belt as boundaries of south and north, and converges westwards and links Qilian sea, as well as opens eastwards. The Ordos area belongs to western section of the basin, and the pattern of stable continental plate against movement belt of continental margin trends toward union. During late period of Late Palaeozoic, sedimentary basin of intraplate epicontinental sea is turned to continental deposit, and this depositional – tectonic pattern is basically inherited during Triassic. Western and southwestern margin of the area come to uplift intensively because of ocean crust subduction within Tethys tectonic domain, and more than 3000m in thickness, coarse clastic sediments are formed in front of the margin. At this time, Ordos Basin belongs to a part of huge intracontinental basin, which takes Caledonian folded belt as depositional boundaries of south and north, and western side of the area features well – developed thrust fold. A systematic discussion on the huge intracontinental basin is made by Liu Shaolong(1986), and he considered that there exist an united giant basin, range close to $9 \times 10^5 Km^2$, in North China at that time, Zhao Chongyuan(1990) considered that the giant basin is a result of convergencing westwards during early period. In late period of Late Triassic, development history of the basin, influenced by Indo – China movement, came to end by means of uplift and denuding.

Early – middle Jurassic to Early Cretaceous, a formation and development stage of Ordos Basin
After Indo – China movement, Chinese geotectonics went into a new period of tectonic development, that is to say, an intensively active period of marginal – Pacific tectonic domain and Tethys tectonic domain. Under compressional stress derived from Pacific direction, a left – lateral shear close to SN trending, and right – lateral shear close to SN trending from WS direction, Ordos area is in a down – warped stage of relative stablity, and results in formation of Ordos Basin. By way of analysis of lithologic characteristics and thickness of Jurassic and Cretaceous in western margin of Ordos, it seems that formation and evolution of the basin is considerably influenced by Tethys tectonic domain. It can be recognized that the basin nowadays is a result of succeeding reform based on analysis of distributed characteristics of strata during Jurassic and Cretaceous, and that initial depositional scope is more than that of the present. (Fig. 1)

SCOPE AND BOUNDARY OF THE BASIN

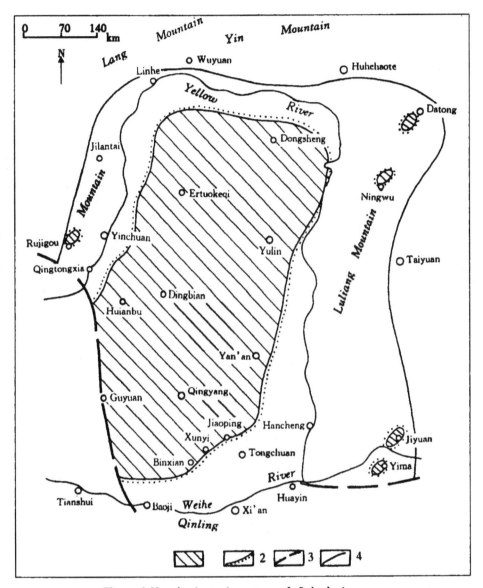

Figure. 1 Map showing various scope of Ordos basin
1. Nowadays; 2. Boundary of outcrop;
3. Fault   4. Boundary of initial deposition

Northern boundary of the basin

Northern margin of Ordos Basin is neighbouring Hetao fault depression. A great amount of exploration have been carried out by Changqing Bureau of Petroleum Exploration in the fault depression. Up to now, it has been found that lamellibranch fossil of late Jurassic to Early Cretaceous exists in Linhe, Huhe and Jilantai depression. By virtue of the discovery, it cannot be denied that there exist Lisangou Fm. and Guyang Fm. . No geological proof that is convinced has been found yet, although some one takes the down – warped depression above mentioned as an alone basin. According to the fact that strata of Middle – Lower Jurassic within northern section of Ordos Basin is overlaped and changed coarse – grained northwards, northern boundary of the basin is approximate to Ula Mts. to Daqing Mts.

Southern boundary of the basin

Southern margin of the basin is neighbouring Weihe fault depression. Due to the fact that there is no drill hole through Cenozoic Erathem in Weihe fault depression, southern boundary of the basin is only determinated by means of analysis of earthquake exploration data here. It has been illustrated by the data that there is obvious difference in basement tectonic at both sides of the Weihe fault, that is, south side belongs to a part of Qiling orogeny belt, and north side to a part of Weihe uplift. By way of analysis of the fact that strata of Early – Middle Jurassic in Binxian, Xunyi and Jiaoping is decreasing in thickness, coarse – grained and poorly – sorted lithologically, southern margin of the basin might be limited to north of the Weihe fault.

Western boundary of the basin

Western margin of the basin is characterized by strong change of later tectonics, So, there are arguments about determination of the western margin, and point at issue is the ownership of Rujigou coal field. In Rujigou coal field, not only lithologic characteristics, depositional environment and underlay strata of Jurassic can be compared with inside of the basin all about, but also palaeoflow direction at both sides of synclinal points at all to centre of Ordos Basin. It is showed that sedimentary strata of Rujigou, Early to middle Jurassic, belongs to a part of Ordos Basin, and western boundary might be located at the western foot of Helan Mountains to Qingtongxia and Guyuan.

Eastern boundary of the basin

Most of Jurassic strata within the eastern basin have been denuded, Yan' an area, which is mainly composed of lacustrine sediments except for sediments of stream system in bottom, and, obviously, it is not boundary of depositional basin. During Jurassic, Luliang Mountains has not been formed based on analysis of regional tectonics. At that time, depositional system around the basin opens eastwards, and centripetal pattern is not formed according to analysis[2] (Fig. 2). By virtue of coal – bearing features of Yan' an Fm. seam are symmetrically developed on both sides of the basin centre within depositional section through the basin SN trending(Fig. 3). But, in depositional section, EW trending(Fig. 4), seam is only developed on west side of the basin centre.

It has been illustrated above that coal – bearing strata of Jurassic in Datong and Ningwu, east of Luliang Mountains together with Jurassic sediments of Jiyuan, Henan Province might be a part of Ordos Basin. Coal – bearing strata of Jurassic, from Datong to Ningwu, transites from clastic coal – bearing strata into lacustrine sediments that are poor coal – bearing or without coal. Those characteristics are similar to that of Shenmu to Yan' an, west of Luliang Mountains. Deep – seated lacustrine facies and sediments of profundal gravity flow are found in Jurassic strata of Jiyuan. They might take together with lacustrine sediment nearby Yan' an to form a subsidence centre of the basin[5], On that score, eastern boundary of Ordos Basin might be east of Datong and Yima, Shanxi Province.

EVOLUTION AND DEVELOPMENT OF THE BASIN

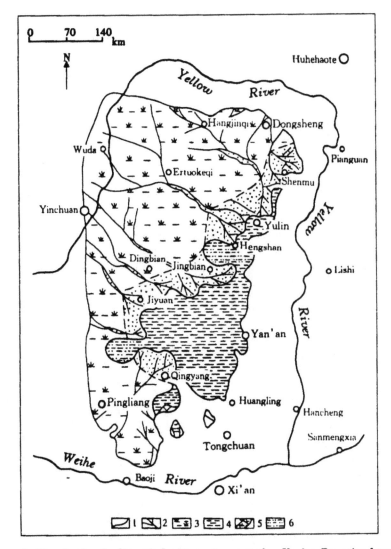

Figure. 2 Map showing the disposal of palaeoenvironment when Yan'an Formation formed

1. Boundary of deposition and denuding; 2. Fluvial course;
3. Fluvial plain; 4. Open lacustrine; 5. Delta; 6. Lagoon

Ordos Basin, marked by bearing obvious stages as respects to formation and evolution, can be divided into four epoches based on geological boundary, types and sedimentary thickness as well as location of depositional centre.

First epoch occurs after Indo–China movement, and represented by unconformity before 1st episode of Yanshan movement. Coal–bearing sediments is formed during late period of Early Jurassic to early period of Middle Jurassic. In the epoch, depositional systems of alluvial, delta and lake are well–developed within the basin. Sediments in stratigraphically ascending order can be divided into three stages: initial, overlapping and offlapping based on boundary of main coal seams representing extensive inorganic

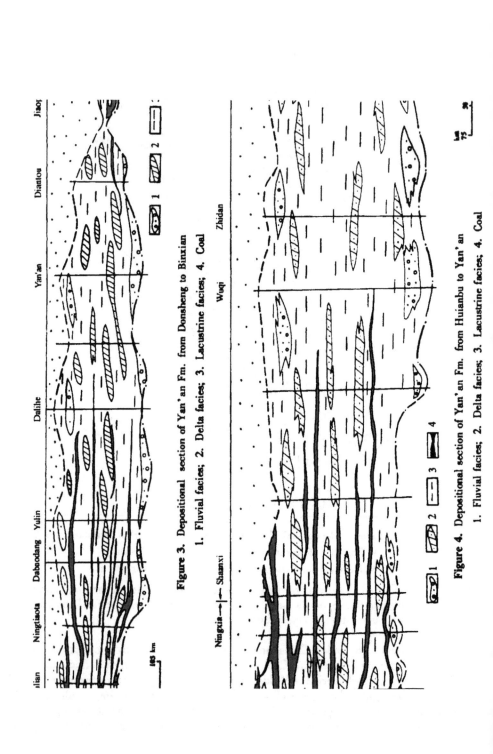

Figure 3. Depositional section of Yan'an Fm. from Donsheng to Binxian

1. Fluvial facies; 2. Delta facies; 3. Lacustrine facies; 4. Coal

Figure 4. Depositional section of Yan'an Fm. from Huianbu to Yan'an

1. Fluvial facies; 2. Delta facies; 3. Lacustrine facies; 4. Coal

abandonment, and maxmum lake expansion surface as well as plane matching features of three depositional systems above mentioned.

In the early stage of initial filling, the morphology of the basin is uneven and complex, sediments are obviously controlled by rolling morphology of paleo – structural plane in Indo – China movement. Sandy – argillaceous sidiments within shallow lake are mainly developed in paleo – swales separated each other, and sandy – gravelly fluvial sediments are mainly developed in paleo – valley. There is not deposition in paleo – highland. Its sediments are characterized by close material source, short – range transportation and fast accumulation. Along with paleo – swale being filled, confined lakes separated each other are linked up, forming a large united confined lake.

In the overlapping filling stage, the basin entered a new deposition configuration of fluvial, delta and lake jointly action. Lake is developed to the utmost extent except for a few districts of the basin margin, and palaeo – highland disappears. Fluvial and delta systems have been shrunk to margin area of the basin, and lacustrine centre is situated at Yan'an and east of Yan'an. An accumulation pattern of strata is typically presented in the form of overlapping. The rock type is mainly sandstone(41. 97%), secondly argillite (26. 81%)and siltstone (24. 78%).

In the offlapping filling stage, lake is shrunk towards centre of the basin, and fluvial and delta deposition is advanced towards centre of the basin. Thus, a typical progradational accumulating pattern is formed, in which mainly is sandstone(51. 62%), secondly siltstone (21. 85%)and argillite (21%).

The drill hole data showed that variation of strata thickness in range of $200000Km^2$ is limited in numbers. In general, the variation range is 180 to 300m(Fig. 5), and mean accumulation velocity is about 20m/Ma. This is a good response to structural condition of relative stablity, and to slow subsidence of earth crust. Depositional centre is situated at Lingwu to Weizhou based on strata thickness delineation, and depression direction is NE 35°.

Depositional strata of the second epoch are Zhiluo and Anding Fm. which are formed during middle and late period of Middle Jurassic. Due to the fact that structural variation of the first episode of Yanshan movement within the basin is very weak, and its basic structural plane combined with scope are not essentially changed. During middle period of middle Jurassic, the basin is all filled with sediments of fluvial depositional system. In late period, the basin is mainly filled with sediments of fresh lacustrine system, and sediments of delta system is only developed in margin of the basin. Red clastic rock series are widespreadly developed, as influenced by palaeo – climate, and coal – accumulating process disappears. Total thickness of the second epoch sedimentary strata is 240 to 650m, and mean accumulation velocity is 12m/Ma. Depositional centre is situated at Huanxian based on stratigraphic isopach, and depression direction is NE25°(Fig. 6).

Structural variation, which occurs in the second episode of Yanshan movement during late period of Middle Jurassic, resulted in withering and disappearance of the second epoch of the basin. During Late Jurassic, sedimentary basin of the third epoch is formed and is basically under compressional stress just like other basin in eastern area of China. Depositional basin of the third epoch, represented by Fengfanghe Formation, is

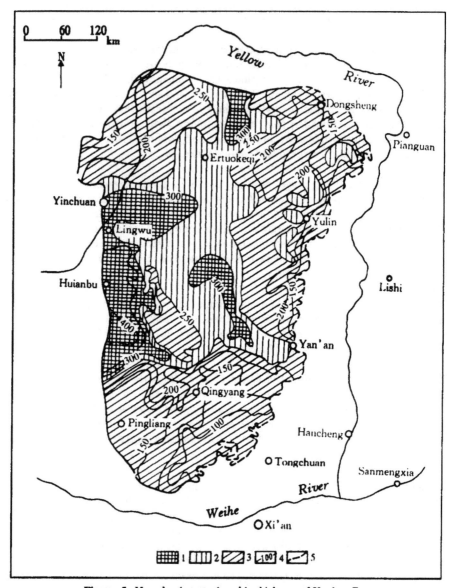

Figure. 5 Map showing stratigraphic thickness of Yan' an Fm.

1. >300m;   2. 250 to 300m;   3. <250m;

4. Isopach of thickness;   5. Outcrop line

converged westwards into a banding depression along east foot of Zhuozi Mountains to Pingluo and Qianyang. Depositional strata derived from paroxysmal talus of the west structurally active area. Those talus are composed of wedges, a kind of red conglomerate, different in thickness at SN strike. Its maximum thickness is 2000 m, minimum 109 m, and angular unconformity with underlying strata is obvious. This showed that the third epoch of Ordos Basin with respects of structure is not quiet when it formed.

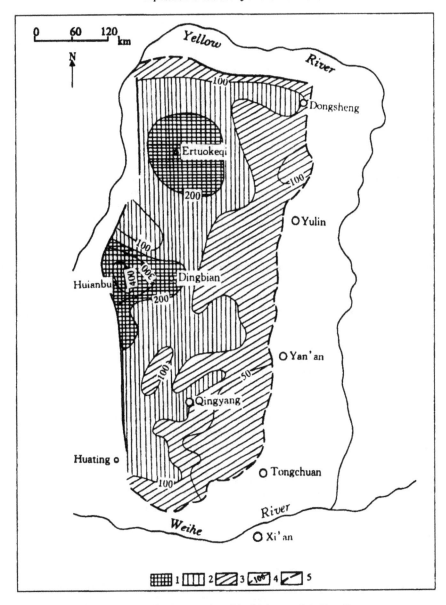

Figure. 6 Map showing stratigraphic thickness of Anding Fm.
1. >200m; 2. 100 to 200m; 3. <200m;
4. Isopach of thickness; 5. Outcrop line

Depositional basin of the fourth epoch is formed during initial period of Cretaceous, influenced by principal episode of Yanshan movement. Just like the third epoch basin, settlement is obviously developed in fore margin of thrusting and napping systems within margin of the basin. Gravelly talus, derived from structurally active region, is laterally filled in the basin with SN stretching. At the same time, superposition and overlapping of strata took place, expanded eastwards, and its eastern boundary might be west to Yellow

River. After development stage of fluvial plain, sediments of lacustrine depositional system are filled. Total thickness of sedimentary strata in the fourth epoch is 600 to 1200m, and mean accumulation velocity is 33m/Ma. Depositional centre is situated west to Huanxian, and depression direction is close to SN (Fig. 7).

During late period of Early Cretaceous, Ordos Basin is finally withered away, and then whole area is a in state of slow uplift.

PROCESS AND REGULARITY OF COAL – ACCUMULATION

The basin bears three units of coal – bearing strata, that is, Carboniferous and Permian, Triassic and Early – Middle of Jurassic, and total reserve of coal is close to 2×10^{12} tons. It is an exceptionally large coal – bearing basin in the world. In fact, coal – bearing strata of Carboniferous, Permian and Jurassic exist before Ordos Basin is formed[3]. The present paper does not deals in detail with the process and regularity of coal – accumulation before the basin is formed. General background of Jurassic coal – accumulation is a large inland basin. In stratigraphically ascending order, five units of seam in coal – bearing strata during Early to Middle Jurassic are formed within the basin, and except that there is not seam accumulation in central basin where is lacustrine deposition for long, that is, Yan' an, Yanchang and Yanchuan (Fig. 8). The transitional period between filling stages is the best period for coal – forming. The transitional surface between different depositional systems is a favourable location for coal formation. the above two situations are the results of abandonment of inorganic sedimentation and superiority of organic sedimentation[6].

The overall distribution of the seam is controlled by morphology of the depression, and coal – accumulating area around subsidence centre of the basin where is seamless, takes the form of ring. The number, thickness and laterally changed regularity of the seam vary with different positions of the basin. Only one major minable seam is developed in the south basin, and situated at the lowest position of coal – bearing strata. Plane distrubution scope, thickness and structure of the seam are obviously controlled by palaeotopography before coal formation. It can be recognized that the seam, with maximum thickness, most simple in structure and being lowest in ash, is situated at the axis of palaeodepression, and, thinning and pinching out towards palaeo – highland (Fig. 9)

World – famous mining area of Shenfu, is justly situated at the northern basin. Here five units of seam are well – developed. Major minable seam with maximum thickness, is located on upper of coal – bearing strata. and characterized by widespread distribution, stable thickness and low ash. The seam is diverged and pinched out when thick – coal area transited towards margin and centre of the basin[4, 6]. The west basin is characterized by lots of seam, small thickness of single layer and diverging combined with merging.

MAJOR FACTORS CONTROLLED ACCUMULATION OF SEAM

There are various factors controlled accumulation of seam, e. g. palaeoclimate, palaeo – tectonics and intensity of organic sedimentation, with regard to overall features of Ordos Basin.

Palaeoclimate, a key factor controlled seam formation, provided environment in favour of plant growth and breeding. It is usually considered by documents that the moist and warm climate is favourable for coal formation, and determination of the properties of palaeolclimate is based on the feature of plant community. But, it is showed by sporopollen data that there is a great number of *Classopollis*, represented drying temperate of subtropical zone by its primordial plant, in Fuxian Formation and Yan'an Formation. Combined with analysis of formation and distribution of modern Earth's surface peat, the authors con-

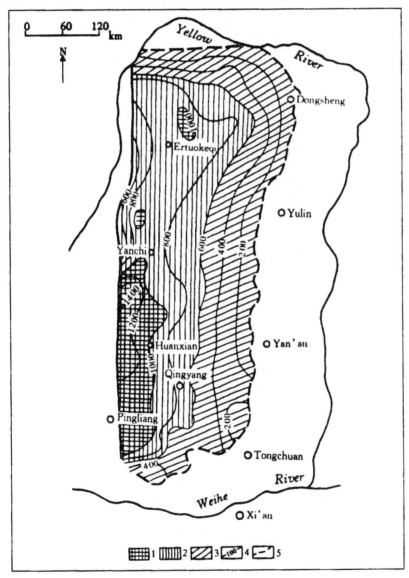

Figure. 7 Map showing stratigraphic thickness of Zhidan Group

1. >1000m; 2. 600 to 1000m; 3. <600m;

4. Isopach of thickness; 5. Outcrop line

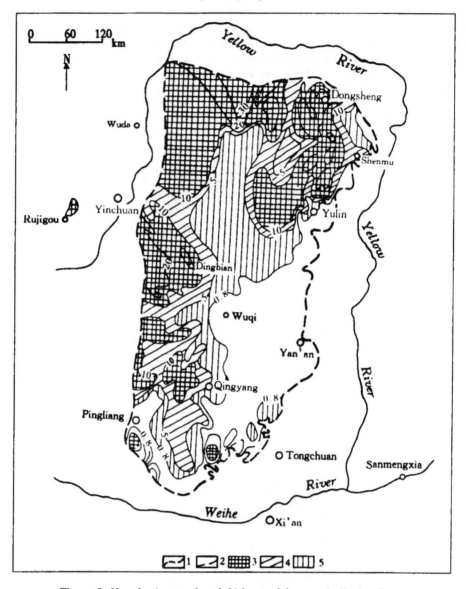

Figure. 8 Map showing grand total thickness of the seam in Yan' an Fm.

1. Boundary preserved rock series of coal – bearing;
2. Isopach; 3. >10m; 4. 10 to 5m; 5 5 to 0. 8m

sider that climate, when coal – bearing strata formed during Early to Middle Jurassic within Ordos Basin, belongs to an alternative type which is warm and moist climate in favour of plant rapid growth together with drying and cold climate resulted in dying out of plant. The judge above mentioned is supported by widespread distribution of thin bed inertinite in coal.

It is showed by space – time distribution features of seam in Ordos Basin that the seam is

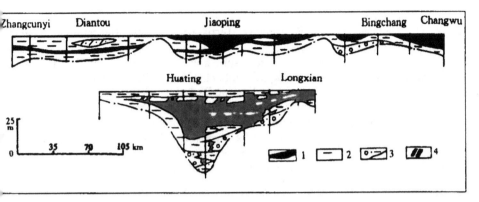

Figure 9. Deposition sectional drawing of No. 5 coal unit in Yan'an Fm. South margin of the basin
1. Seam; 2. Shallow lacutrine facies; 3. Fan – shaped facies; 4. Oil shale

controlled by tectonics which displays in two respects, that is, coal – accumulating period and coal – accumulating area. In detail, the coal – accumulating period is usually controlled by structural transitional period, represented by coal – bearing strata of Early to Middle Jurassic, which is formed in transitional process from Indo – China with lift to Early – Yanshan with subsidence. Also, five major minable seams in Yan' an Fm. which bears very similar features to that above mentioned, that is to say, fluvial and delta depositional system, marked by lift tectonic setting, are usually under the seam; lacustrine depositional system, marked by subsidence setting, is usually over the seam. Coal – accumulating area is usually controlled by tectonic transitional position. For example, overall plane morphology of seam in Yan'an Fm. is in the form of ring, and the ring is in turning position from lift area of the basin margin to subsidence area of the basin centre. In the process of formation and evolution of Ordos Basin, southwest margin of the basin is continuously in tectonic active area. Just in the area, seam thickness is in maximum, represented by Huating mining area.

The intensity of organic sedimentation is controlled by palaeoclimate and geotectonic setting. The palaeoclimate can produce a great amount of plant remains, and provide organic sedimentation with material. The geotectonic setting can result in abandonment of inorganic sedimentation, and provide transportation and deposition of the plant remains with condition. This tectonic setting only occures in tectonic turning period. Such being the case, perfect match of palaeoclimate with tectonic turning period is a key factor determined intensity of organic sedimentation, and the intensity is also a key factor determined coal forming or not , and determined inorganic contents in coal[6]. On that score, it is reputed that the effective ways of strategically coal – hunting might be palaeoclimate restoration, by means of palaeomagnetic research, combined with tectonic settings analysis.

REFERENCE

1. W. R. Dickinson, Plate tectonics and sedimentation, Tectonics and sedimention, Soc.

Econ. Palarontol Mineral, Spec. publ. 22(1974).

2. W. E. Galloway & D. K. Hobday, Terrigenous Clastic Depositional System, Springer Verlay(1983).

3. Li Shitian, et al. , Sequence Stratigraphy and Deposition System Analysis of the Northeastern Ordos Basin, Geological Publishing House, Beijing(1992).

4. Shaanxi Coal Geological Bureau, Early – middle Jurassic Sedimentary Enviroment of the Northern Shaanxi, Shaanxi Science and Technology Publishing House(1989).

5. J. C. Van Wagoner, R. M. Mitchum, et al. , Siliciclastic Sequence Stratigraphy in Well Logs, Cores and Outcrops: Concept for high – Resolution Correlaiton of Time and Facies, AAPG Mothods in Exploration Series, No. 7.

6. Wang Shuangming et al. , Coal Accumulation and Coal Resource Evaluation of Ordos Basin, China Coal Industry Publishing House, Beijing (1996).

Proc. 30th Int'l Geol. Congr., Vol. 18, Part B, pp. 47-57
Yang Qi (Ed.)
© VSP 1997

Geologic Factors Affecting the Abundance, Distribution, and Speciation of Sulfur in Coals

C.-L. CHOU
Illinois State Geological Survey, 615 East Peabody Drive, Champaign, IL 61820, U.S.A.

Abstract

Sulfur in coals is derived partly from original plant materials and partly from ambient fluids in the coal-forming environment. Low-sulfur (LS) coal seams (less than 1% total sulfur content), such as Tertiary coals of the Powder River Basin, USA, were deposited in an alluvial environment and the peat was not influenced by seawater. High-sulfur (HS) coal seams (greater than 3% total sulfur content) are usually associated with marine strata. Seawater influence was the primary cause of the high sulfur in the coal. For example, in the Illinois Basin, high-sulfur Herrin Coal is overlain by marine Anna Shale and Brereton Limestone, but, medium-sulfur (MS) coal (1%-3% total sulfur content) occurs in well-defined areas of the Herrin seam where the coal is directly overlain by thick (>6 m) Energy Shale. This thick nonmarine gray shale was an effective barrier to seawater diffusion into peat. Superhigh-organic-sulfur (SHOS) coals have been reported, including Rasa coal from Croatia, which has a total sulfur content of 10.8% and organic sulfur content of 10.5% (dry basis). A SHOS coal of Late Permian age from Guidin, Guizhou, China contains 9.2-10.5% total sulfur content and 8.2-9.2% organic sulfur content (dry basis). Guidin coal was derived from peat deposited on a coastal carbonate platform. Profound seawater influence and iron deficiency are key factors in the formation of SHOS Guidin coal. The sulfur content of coal macerals is highly variable. Sporinite embedded in vitrinite is higher in organic sulfur than surrounding vitrinite in Illinois Basin coals. A high efficiency of organic sulfur formation in sporinite may be resulted from abundant functional groups in sporinite precursors which reacted with active sulfur species (hydrogen sulfide, elemental sulfur and polysulfides) during early diagenesis. The thiophenic fraction of organic sulfur increases with increasing carbon content of coal. It is interpreted that organic sulfur species initially formed during early diagenesis are mostly thiols and sulfides, which convert to thiophenic compounds during coal maturation. Thus the thiophenic fraction of organic sulfur increases with coal rank.

Keywords: coal, sulfur in coal, organic sulfur species, superhigh-organic-sulfur coal, thiophene

INTRODUCTION

The knowledge about the abundance and nature of sulfur in coal is important in coal utilization because sulfur oxides released during coal combustion can be a major cause of acid rain. Back in 1990 I reviewed the geochemistry of sulfur in coal [12]. It was concluded then that the variation of sulfur in coal is controlled by geologic conditions, and sulfur in plant material is the principle source of sulfur in low-sulfur coal. In medium and high-sulfur coals, most of the sulfur is derived from seawater sulfate; the formation of medium- and high-sulfur coal is a result of seawater influence on peat during early diagenesis. Recent studies showed that superhigh-organic sulfur coals occur in special sedimentary environments in which organic matter is profoundly influenced by seawater. This paper discusses the relation between the sulfur content in coal and sedimentary environments of coal formation [13].

Further, the incorporation of organic sulfur in coal during maceral formation involves reactions of reduced sulfur species (hydrogen sulfide, elemental sulfur, and polysulfides) with organic matter [12]. Such reactions may be more efficient for certain maceral precursors which are abundant in reactive functional groups. This may have controlled the relative organic sulfur abundances in macerals of each coal. It is likely that organic sulfur initially formed in coal macerals during humification and gelification stages is mostly thiols and sulfides. These sulfur species should evolve during various stages of coalification, i.e. thiols and sulfides convert to thiophenic compounds [13, 14]. The relation between the abundance of thiophenic fraction of organic sulfur and carbon content of coal is examined, and its implications on the sulfur evolution during coal maturation will be discussed [13,14].

SULFUR ABUNDANCE AND DEPOSITIONAL ENVIRONMENT OF COAL

The level of sulfur content in coals has been divided into low-sulfur (LS) coal that contains less than 1% total sulfur, medium-sulfur (MS) coal that contains 1% to <3% total sulfur, and high-sulfur (HS) coal that contains ≥3% total sulfur [12].

Low-sulfur coals, such as Tertiary coals of the Powder River Basin, USA, were deposited in an alluvial environment, in which the coal had never been influenced by seawater during the entire geologic history. Depositional models of Tertiary and Cretaceous coal-bearing fluvial deposits in Powder River and other basins in the Rocky Mountain region of the U.S. were developed [20-22].

High-sulfur (HS) coal seams (greater than 3% total sulfur content) are usually associated with marine strata. Seawater influence was the primary cause of the high sulfur in the coal. For example, in the Illinois Basin, high-sulfur Herrin Coal is overlain by marine Anna Shale and Brereton Limestone, but, medium-sulfur (MS) coal (1-3% total sulfur content) occurs in well-defined areas of the Herrin seam where the coal is directly overlain by thick (>6 m) Energy Shale, a gray-shale deposit along a contemporary river. This thick non-marine shale was an effective barrier to seawater diffusion into peat [11].

Kvale and Archer [32] recently recognized tidal deposits associated with low-medium sulfur coal deposits in the Lower and Upper Block Coal Members of the Brazil Formation (Lower Pennsylvanian) in Indiana, USA. The Lower Block Coal has a sulfur content of 0.7%-1.8%, averaging 1.2% [37, 38]. The silty rhythmites are suggested to be deposited in an estuarine environment in brackish to nearly freshwater settings [2].

The Energy Shale was commonly interpreted as freshwater crevasse-splay and lacustrine deposits [7, 19]. Recently, Archer and Kvale [3], based on their sedimentological study, reinterpreted all gray shales above low-medium sulfur coals in the Illinois Basin as tidally influenced estuarine/deltaic deposits.

The relation between sulfur content and sedimentary environment was also independently observed for Chinese coal seams of Permo-Carboniferous, Jurassic, Cretaceous and Tertiary ages. Zhao [46] showed that Chinese coals formed in two broadly different categories of sedimentary environments have different sulfur content. Coal seams that occur in an alternating marine-terrestrial facies (nearshore, littoral environment) generally have a high sulfur content (1-5% sulfur), whereas coals that occur in an transitional (delta plain) and terrestrial facies (inland fluvial or lacustrine environment have a low sulfur content (<1-2% sulfur). Therefore the sulfur content in coal is basically controlled by sedimentary environment of coal formation.

Table 1. Superhigh-organic-sulfur (SHOS) coals and their total sulfur content (TS), sulfur forms and depositional environment (concentrations on a moisture-free basis)

Coal	Sulf S	Pyr S	Org S	TS	Ref.
Rasa, Istrian Peninsula, Croatia [roofs of coal beds are mainly marine limestone and flysch]					
	--	--	--	10.4	[31]
	0.02	0.30	10.52	10.84	[45]
Guidin, Guangxi, China [tidal flat environment of limited carbonate platform]					
GD-1	0.05	1.14	8.30	9.49	[36]
GD-2	0.11	0.82	8.23	9.16	[36]
GD-3	0.23	1.05	9.18	10.46	[36]
K3	0.16	1.02	7.71	8.89	[39]
Yanshan, Yunnan, China [tidal flat environment of limited carbonate platform]					
M7-9	0.20	0.80	10.30	11.30	[39]
Wuyi, Anxian, Sichuan, China [floor is residual plain, and roof is tidal flat environment of limited marine carbonate platform]					
TLZ	0.08	3.56	7.58	11.13	[40]
DLZ	0.16	5.97	7.01	13.14	[39]

ORIGIN OF SUPERHIGH-ORGANIC-SULFUR COAL

Coal that is greatly enriched in organic sulfur is called superhigh-organic-sulfur (SHOS) coals. Table 1 is a compilation of total sulfur and sulfur forms in the SHOS coals. All SHOS coals show extreme enrichment of organic sulfur, from 7.0% to 10.5%, but the pyritic sulfur content is variable (Fig. 1). Rasa, Yanshan and Guidin coals have a low pyritic sulfur content of 0.30% to 1.1%. A low pyrite content in these coals is because the supply of iron is limited in a coal forming environment of tidal flat of limited carbonate platform. The sulfur isotopic composition of Guidin coal (pyritic sulfur with $\delta^{34}S$ values of -28.2 to -30.6‰, sulfate sulfur -22.7 to -30.6‰, and organic sulfur -7.4 to -7.7‰) indicates that bacterial reduction of seawater sulfate was an important process in the formation of organic and pyritic sulfur [36].

Pyritic sulfur content in two samples of Wuyi coal are 3.56% and 5.97%, respectively. A high pyrite content in the Wuyi indicates that there is a significant supply of iron in its depositional setting. The floor of coal seam is a residual plain. Thus, the SHOS coals can be low or high in pyrite content depending on the level of iron supply in their depositional settings.

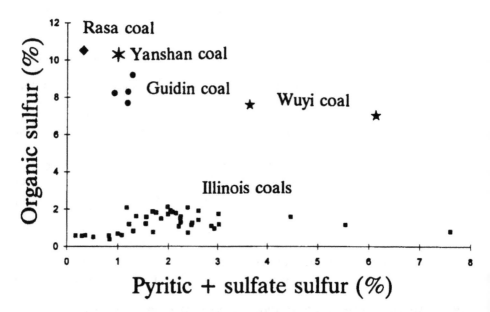

Figure 1. Organic sulfur versus pyritic+sulfate sulfur contents in superhigh organic sulfur (SHOS) coals as compared with Illinois Basin, USA, coals. Rasa, Yanshan, and Guidin coals are low-pyrite SHOS coals, and Wuyi coal is high-pyrite SHOS coal. The data are taken from the literature [25, 31, 34-36, 39, 40, 45].

VARIATION OF ORGANIC SULFUR IN COAL MACERALS: IMPLICATIONS ON ORGANIC SULFUR FORMATION

The methods for determination of organic sulfur in coals were reviewed by Chou [12], Davidson [16], and Calkins [10]. The distribution of organic sulfur was determined in coal macerals separated by density gradient centrifugation using a transmission electron microscopic (TEM) technique [42, 44], and in macerals of Illinois Basin coals using scanning electron microscope (SEM) in conjunction with energy dispersive X-ray spectrometer (EDX) [17]. The organic sulfur content in maceral groups decreases in the order of exinite, vitrinite, and inertinite. The exinite group has a large range of organic sulfur content perhaps because of the presence of different macerals in the group. Hippo et al. [27] showed that sporinite is higher in organic sulfur than other macerals in the exinite group. The spatial distribution of organic sulfur by TEM and SEM-EDX traverses showed that a sporinite embedded in vitrinite is higher in organic sulfur content than the adjoining vitrinite, and organic sulfur is higher in vitrinite than adjoining semifusinite [17, 42, 44].

The systematic variation of organic sulfur content in coal maceral groups of each coal (high in sporinite, intermediate in vitrinite, and low in inertinite) indicates that the efficiency of organic sulfur formation during early diagenesis decreases in this order. It was noticed earlier that the H/C ratio of macerals decreases in the same order of sporinite, vitrinite, and inertinite [17, 41].

In a reducing environment of coal diagenesis, hydrogen sulfide is produced by bacterial reduction of seawater sulfate. Sulfate itself is considered nonreactive under the conditions of peat formation. Polysulfides, an oxidation intermediate of hydrogen sulfide, play an important role in incorporating sulfur into organic matter [e.g., 33]. Elemental sulfur (S_8) cannot react with organic compounds directly, but it may be involved indirectly through the formation of polysulfides [1]. The processes of organic sulfur formation involves the reaction of reactive sulfur species (hydrogen sulfide, elemental sulfur and polysulfides) with organic matter during humification and gelification stages of maceral formation. The formation of organic sulfur appears to be most efficient for sporinite precursor, which is the richest in reactive function groups, and the least efficient in fusinite precursor because fusinite loses most of reactive functional groups by oxidation during fusinite formation.

THE EVOLUTION OF ORGANIC SULFUR SPECIES DURING COAL MATURATION

Organic sulfur species in coals are mainly thiols, sulfides, disulfides, and thiophene and its derivatives [12]. Calkins [8] suggested that the actual aliphatic sulfur structures in coal are thioethers. Ignasiak et al. [29] reported that the presence of thioethers and a lack of thiols in Rasa coal.

In earlier work on chemical structure of different ranks of coals, the sulfur-

containing aromatic compounds (thiophene and its derivatives) were found in
bituminous coal and anthracite, but not in lignite [26]. Attar and coworkers [4-6]
developed a thermokinetic method and estimated that 40% to 60% of the organic
sulfur in bituminous coal is thiophenic compounds, and the rest is thiolic and sulfide
compounds. Chou [12, p. 39] suggested that the relative abundance of various types
of organic sulfur compounds may be related to the rank of coal.

Pyroprobe GC-MS technique was used by Calkins [8] to determine the abundance
of organic sulfur species in a series of coal with a range of Btu values, and the
results showed that the thiophenic fraction of organic sulfur increases with
increasing Btu values of coals. Recently, two X-ray techniques were developed for
determining organic sulfur species in coal: an X-ray absorption near-edge structure
spectroscopy (XANES) and X-ray photoelectron spectroscopy (XPS) [23, 28, 30].
The abundance of organic sulfur species was determined using the XANES
technique for eight Argonne Premium Coal Samples (varying from lignite to low
volatile bituminous coal in rank) [24], a suite of Illinois Basin coal samples (high
volatile bituminous coals) [15], and superhigh-organic-sulfur coals (Rasa coal from
Istria, Croatia; Charming Creek coal, New Zealand; and Mequinenza coal, Spain)
[24].

The relation between the thiophenic fraction of organic sulfur and carbon content
in these coals is shown in Fig. 2. The thiophenic fraction of organic sulfur increases
with increasing carbon content, a coal rank parameter. The thiophenic fraction
constitutes about 52-70% of organic sulfur in lignite and subbituminous coals. When
the rank increases to high volatile bituminous coals, such as coals from Illinois
Basin, Pittsburgh seam of Pennsylvania, Blind Canyon seam of Utah, and Lewiston-
Stockton seam of West Virginia, the thiophenic fraction reaches 62-82% of organic
sulfur [15, 24, 43]. In the medium volatile bituminous coal from Upper Freeport
seam, Pennsylvania, and low volatile bituminous coal from Pocahontas seam,
Virginia, 87% of organic sulfur is in the form of thiophenic compounds [24, 43].

Hence a sequence of organic sulfur evolution in coals is established. During early
diagenesis, organic sulfur compounds formed in peat are mostly thiols and sulfides.
These sulfur compounds gradually convert to thiophenes during coal maturation.
The thiophenic fraction thus increases from lignite, through subbituminous coal to
bituminous coal. The thiophenic fraction of organic sulfur would presumably be the
highest in anthracite.

CONCLUSIONS

This paper summarizes significant advances in our understanding of the origin and
evolution of sulfur in coals. Based on sulfur content, there are low- (LS), medium-
(MS), high-sulfur (HS), and superhigh-organic sulfur (SHOS) coals. The abundance
of total sulfur and sulfur forms are controlled by depositional environments of coal
formation. Sulfur content in coal increases with increasing degree of marine

influence. Low-pyrite SHOS coals were formed in a tidal flat environment of carbonate platform in which the organic matter was profoundly influenced by seawater. The depositional environment is characterized by an ample supply of sulfur but a scarce source of iron.

The organic sulfur content in macerals decreases in the order of sporinite, vitrinites, and fusinite. Sporinite precursor appears to be most abundant in functional groups which reacted with reduced sulfur species generated by bacterial reduction of seawater sulfate to form organic sulfur. The organic sulfur species formed during early diagenesis are mostly thiols and sulfides. As coalification processes progress, the non-thiophenic sulfur species convert to thermally stable thiophenes. Thus, similar to vitrinite reflectance and aromaticity, the thiophenic fraction of organic sulfur increases with the degree of thermal maturity of coal.

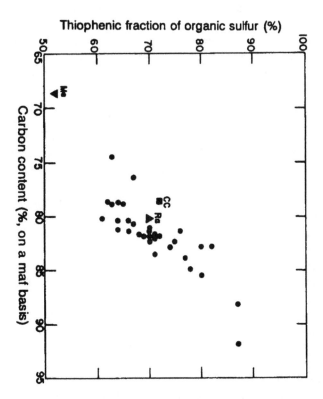

Figure 2. Relation between percentage thiophenic fraction of organic sulfur and carbon content (on a moisture and ash-free basis) in coals. The rank of coals ranges from lignite, subbituminois coal, and high, medium and low volatile bituminous coals. All samples are U.S. coals except Rasa coal from Istria, Croatia (Ra), Charming Creek coal from New Zealand (CC) and Mequinenza coal from Spain (Me). The thiophene abundance data are those determined by the XANES technique. The thiophenic sulfur and carbon content data are taken from the literature [9, 15, 18, 24, 43].

Acknowledgments

I thank Professor Ren Deyi for inviting me to visit the Beijing Graduate School, China University of Mining and Technology in August, 1996. I thank Professor Ren, Professor Zhang Pengfei, Dr. Lei Jiajin, and Dr. Tang Yuegang for helpful discussions on the topic of sulfur in coals.

REFERENCES

1. Aizenshtat, E.B. Krein, M.A. Vairavamurthy and T.P. Goldstein. Role of sulfur in the transformations of sedimentary organic matter: a mechanistic overview. In: *Geochemical transformations of sedimentary sulfur.* M.A. Vairavamurthy and M.A.A. Schoonen (Eds). Chapter 2, pp. 30-52. Amer. Chem. Soc. Symp. Series **612** (1995).
2. A.W. Archer and E.P. Kvale. Origin of gray-shale lithofacies ("clastic wedges) in U.S. midcontinental coal measures (Pennsylvanian): An alternative explanation. In: *Modern and ancient coal-forming environments.* J.C. Cobb and C.B. Cecil (Eds). pp.181-192. Geol. Soc. Amer. Spec. Paper **286** (1993).
3. A.W. Archer, H.R. Feldman, E.P. Kvale, and W.P. Lanier. Comparison of drier-to wetter-interval estuarine roof facies in the Eastern and Western Interior coal basins, USA. *Palaeogeogr. Palaeoclimatol. Palaeoecol.* **106**, 171-185 (1994).
4. A. Attar. Sulfur groups in coal and their determinations. In: *Analytical methods for coal and coal products.* C. Karr (Ed). III, pp. 585-624. Academic Press, New York (1979).
5. A. Attar and F. Dupuis. Data on the distribution of organic sulfur functional groups in coals. In: *Coal structure.* M.L. Gorbaty and K. Ouchi (Eds). ACS Advances in Coal Chemistry Series 192, 239-256. American Chemical Society, Washington, D.C. (1981).
6. A. Attar and G.G. Hendrickson. Functional groups and heteroatoms in coal. In: *Coal structure.* R.A. Meyers (Ed). pp. 131-198. Academic Press, New York (1982).
7. M.K. Burk, M.P.Deshowitz and J.E. Utgaard. Facies and depositional environments of the Energy Shale Member (Pennsylvanian) and their relationships to low-sulfur coal deposits in southern Illinois. *Jour. Sed. Petrol.* **57**, 1060-1067 (1987).
8. W.H. Calkins. Investigation of organic sulfur-containing structures in coal by flash pyrolysis experiments. *Energy Fuels* **1**, 59-64 (1987)
9. W.H. Calkins, R.T. Torres-Ordonez, B. Jung, M.L. Gorbaty, G.N. George and S.R. Keleman. Comparison of pyrolytic and X-ray spectroscopic methods for determining organic sulfur species in coal. *Energy Fuel* **6**, 411-413 (1992).
10. W.H. Calkins. The chemical forms of sulfur in coal: a review. *Energy Fuels* **73**, 475-484 (1994).
11. C.-L. Chou. Relationship between geochemistry of coal and the nature of strata overlying the Herrin Coal in the Illinois Basin, U.S.A. *Memoir Geol. Soc. China* **6**, 269-280 (1984).

12. C.-L. Chou. Geochemistry of sulfur in coal. In: *Geochemistry of fossil fuels.* W.L. Orr and C.M. White (Eds). Chapter 2, pp. 30-52. Amer. Chem. Soc. Symp. Series **429** (1990).

13. C.-L. Chou. Geologic factors affecting the abundance, distribution, and speciation of sulfur in coal. *30th International Geological Congress Abstracts,* **2,** 877 (1996).

14. C.-L. Chou. Origin of superhigh-organic-sulfur coal and evolution of organic sulfur species during coal maturation: a review. *Geol. Soc. Amer. Abs. Prog.* **28:7,** A-209 (1996).

15. M.-I.M. Chou, I. Demir, R.R. Ruch, J.M. Lytle, S. Bhagwat, Y.C. Li, C.-L. Chou, F.E. Huggins, and G.P. Huffman. *Advanced characterization of forms of chlorine, organic sulfur, and trace elements in available coals from operating Illinois mines.* Final Technical Report, September 1, 1994 through August 31, 1995 to Illinois Clean Coal Institute (1995).

16. R.M. Davidson. Quantifying organic sulfur in coal. *Fuel* **73,** 988-1005 (1994).

17. I. Demir and R.D. Harvey. Variation of organic sulfur in macerals of selected Illinois coals. *Org. Geochem.* **16,** 525-533 (1991)

18. I. Demir, R.D. Harvey, R.R. Ruch, H.H. Damberger, C. Chaven, J.D. Steele, and W.T. Frankie. Characterization of available (marketed) coals from Illinois mines. *Illinois State Geological Survey, Open File Series* **1994-2,** 1-26 (1994).

19. M.J. Edwards, R.L. Langenheim, Jr., W.J. Nelson and C.T. Ledvina. Lithologic patterns in the Energy Shale Member and the origin of "rolls" in the Herrin (No. 6) Coal Member, Pennsylvanian, in the Orient No. 6 Mine, Jefferson County, Illinois. *Jour. Sed. Petrol.* **49,** 1005-1014 (1979).

20. R.M. Flores. Coal depositional models in some Tertiary and Cretaceous coal fields in the U.S. Western Interior. *Org. Geochem.* **1,** 225-235 (1979).

21. R.M. Flores. Styles of coal deposition in Tertiary alluvial deposits, Powder River Basin, Montana and Wyoming. In: *Paleoenvironmental and tectonic controls in coal-forming basins of the United States.* P.C. Lyons and C.L. Rice (Eds). Geol. Soc. Amer. Special Paper **210,** pp. 79-104 (1986).

22. R.M. Flores. Rocky Mountain Tertiary coal-basin models and their applicability to some world basins. *Intern. Jour. Coal Geol.* **12,** 767-798 (1989).

23. M.L. Gorbaty, G.N. George and S.R. Kelemen. Direct determination and quantification of sulphur forms in heavy petroleum and coals. 2. the sulphur K edge X-ray absorption spectroscopy approach. *Fuel* **69,** 945-949 (1990).

24. M.L. Gorbaty, S.R. Kelemen, G.N. George, and P.J. Kwiatek. Characterization and thermal reactivity of oxidized organic sulphur forms in coals. *Fuel* **71,** 1255-1264 (1992).

25. R.D. Harvey, R.A. Cahill, C.-L. Chou and J.D. Steele. Mineral matter and trace elements in the Herrin and Springfield Coals, Illinois Basin Coal Field. *Illinois State Geological Survey, Contract/Grant Report* **1983-4,** 1-162 (1983).

26. R. Hayatsu, R.E. Winans, R.G. Scott, L.P. Moore and M.H. Studier. Trapped organic compounds and aromatic units in coals. *Fuel* **57,** 541-548 (1978).

27. E.J. Hippo, J.C. Crelling, D.P. Sarvela and J. Mukerjee. Organic sulfur distribution and desulfurization of coal maceral fractions. In: *Processing and Utilization of High Sulfur Coals II.* Y.P. Chugh and R.D. Caudle (Eds). pp. 13-

22. Elsevier Science Publishing Company, Inc., New York (1987).

28. G.P. Huffman, F.E. Huggins, S. Mitra, N. Shah, R.J. Pugmire, B. Davis, F.W. Lytle and R.B. Greegor. Investigation of the molecular structure of organic sulfur in coal by XAFS spectroscopy. *Energy Fuels* **3**, 200-205 (1989).

29. B.S. Ignasiak, J.F. Fryer and P. Jadernik. Polymeric structure of coal. 2. Structure and thermoplasticity of sulphur-rich Rasa lignite. *Fuel* **57**, 578-584 (1978).

30. S.R. Kelemen, G.N. George and M.L. Gorbaty. Direct determination and quantification of sulphur forms in heavy petroleum and coals. 1. The X-ray photoelectron spectroscopy (XPS) approach. *Fuel* **69**, 939-944 (1990).

31. D.J.W. Kreulen. Sulphur coal of Istria. *Fuel* **31**, 462-467 (1952).

32. E.P. Kvale and A.W. Archer. Tidal deposits associated with low-sulfur coals, Brazil Fm. (lower Pennsylvanian), Indiana. *Jour. Sed. Petrol.* **60**, 563-574 (1990).

33. R.T. LaLonde. Polysulfide reactions in the formation of organosulfur and other organic compounds in the geosphere. In: *Geochemistry of fossil fuels.* W.L. Orr and C.M. White (Eds). Chapter 4, pp. 68-82. Amer. Chem. Soc. Symp. Series **429** (1990).

34. J. Lei. *Sulfur occurrence regularities of Late Pemian coal in Guizhou province and components, structure and formation of high organosulfur coal.* Ph.D. thesis, Beijing Graduate School, China University of Mining and Technology, Beijing (1993).

35. J. Lei and D. Ren. The petrological and geochemical characteristics of Permian high organosulfur coal from Guidin, China. In: *Processing and utilization of high-sulfur coals V.* B.K. Parekh and J.G. Groppo (Eds). pp. 27-35. Elsevier Science Publishers B.V. (1993).

36. J.-J. Lei, D.-Y. Ren, Y.-G. Tang, X.-L. Chu, and R. Zhao. Sulfur-accumulating model of superhigh organosulfur coal from Guiding, China. *Chinese Science Bulletin* **39**, 1817-1821.

37. P. Padgett, S.M. Rimmer, J.C. Ferm, J.C. Hower, C.F. Eble, M. Thompson, and M. Mastalerz. Sulfur distribution in the Lower Block Coal (southwestern Indiana): inferences about the depositional environment. *Geol. Soc. Amer. Abs. Prog.* **28:7**, A-209 (1996).

38. P. Padgett, S.M. Rimmer, C.F. Eble, J.C. Ferm, J.C. Hower, M. Mastalerz and M. Thompson. Petrology of the Lower Block Coal in southwestern Indiana: implications for the depositional environment. *Thirteenth Ann. Mtg. Soc. Org. Petrol., Abs. Prog.* **13**, 19-21 (1996).

39. D. Ren, Y. Tang, and J. Lei. Study on regularities of sulfur occurrence and pyrite magnetism of Late Permian coals in southwest China. *Jour. China Univ. Mining Tech.* **4:2**, 64-73 (1994).

40. Y. Tang and D. Ren. Research on different pyrites in Late Permian coal of Sichuan province, southwestern China. In: *Processing and utilization of high-sulfur coals V.* B.K. Parekh and J.G. Groppo (Eds). pp. 37-45. Elsevier Science Publishers B.V. (1993).

41. B.P. Tissot and D.H. Welte. *Petroleum formation and occurrence.* Springer-Verlag (1984).

42. B.-H. Tseng, M. Buckentin, K.C. Hsieh, C.A. Wert, and G.R. Dyrkacz. Organic

sulphur in coal macerals. *Fuel* **65**, 385-389 (1986).

43. K.S. Vorres. *Users handbook for the Argonne premium coal sample program.* ANL/PCSP-93/1 (1993).

44. C.A. Wert, K.-C. Hsieh, B.-H. Tseng, and Y.-P. Ge. Applications of transmission electron microscopy to coal chemistry. *Fuel* **66**, 914-920 (1987).

45. C.M. White, L.J. Douglas, R.R. Anderson, C.E. Schmidt, and R.J. Gray. Organosulfur constituents in Rasa coal. In: *Geochemistry of fossil fuels.* W.L. Orr and C.M. White (Eds). Chapter 16, pp. 261-286. Amer. Chem. Soc. Symp. Series **429** (1990).

46. S. Zhao. Reducibility of humic coal in China and its relation to sedimentary environment. *Acta Sed. Sinica* **2:2**, 53-65 (1984)

Proc. 30th Int'l Geol. Congr., Vol. 18, Part B, pp. 59-76
Yang Qi (Ed.)
© VSP 1997

Multistage Metamorphic Evolution and Superimposed Metamorphism through Multithermo-sources in China Coal

YANG QI, WU CHONGLONG, TANG DAZHEN, KANG XIDONG and LIU DAMENG

Department of Energy Resources and Geology, China University of Geosciences, Xue Yuan Road 29, Beijing 100083, P.R.CHINA

Abstract

Through the study of heat source and its operating mode, four types of coal metamorphism are recognized in China. Dynamic simulation of the geologic-thermal history shows that the geothermal state of coal basins influences the geothermal metamorphism of coal in an inhomogeneous geotemperature field. The telemagmatic metamorphism, another significant coal metamorphic type in China, is the main cause that produced a large amount of medium-high rank coals in China. The hydrothermal metamorphism relies mainly on deep circulating hot fluid as heat sources. Contact metamorphism brings about only local effect. Multistage and multithermo-source superimposed metamorphism affects profoundly the evolution of coal metamorphism in China. Coal in the same coalfield may have undergone one or more types of coal metamorphism superimposed on the basically geothermal metamorphism. Large quantity of Chinese coal underwent three evolutional stages: the first evolutional stage is characterized by the predominance of geothermal metamorphism, the second by superimposed metamorphism through multithermo-sources, and the third by the establishment of coal metamorphism frame. Six types of superimposed coal metamorphism are recognized. Take the Fushun basin and the eastern coastal provinces in China as examples, the thermodynamic model for the paleo-geothermal field and its corresponding coal metamorphism is established, and the simulation of two-dimensional multistage and multithermo-source superimposed coal metamorphism is carried out.

The metamorphic regularities of Chinese coal may be briefly summed up like this:the geothermal background of coal metamorphism displays obviously the space-time changes; superimposed metamorphism through multithermo-sources, and especially the telemagmatic metamorphism since the Yanshanian orogeny have evidently reformed the Chinese coal ranks; the Tertiary tectonic displacement brought about the distribution frame of the Chinese coal rank zonation.

Keywords: metamorphism types of Chinese coal, superimposed metamorphism through multithermo-sources and multistages, thermodynamic model, coal metamorphic regularity, China

INTRODUCTION

The continent of China, situated at the southeastern part of the Eurasian plate, is bounded on the north, east and southwest by the Siberian, Pacific and Indian plate respectively. The geotectonic framework of China has become highly complicated due to superimposition and reworking of multiperiodic tectonism. The Indosinian and Yanshanian stages had played important roles on the formation of coal metamorphic framework in China, especially during the Yanshanian stage, the activities of large-scale magmatic intrusives, volcanism and fault blocks had exerted crucial influence on the coal metamorphic types and metamorphic degree of Chinese coals[12-17].

There are four predominant types of coal metamorphism in China:geothermal metamorphism, telemagmatic metamorphism, contact metamorphism and hydrothermal metamorphism. The last type of coal metamorphism is caused by confined water at high temperature from the deeper part of the crust or by differentiated magmatic hydrothermal solution[17].

CHARACTERISTICS OF COAL METAMORPHIC TYPES IN CHINA

Geothermal Metamorphism of Coal

During the Late Paleozoic, the North China and the Yangtze platforms appeared mainly as grand-scaled rolling depressions which constitute then the main coal accumulating areas.Widespread Triassic sediments play an important role in the preservation of the Paleozoic coal measures in East China. During early Yanshanian, there developed steadily younger coal accumulating depressions or superimposed basins mainly of Triassic and Jurassic sequences, such as the Ordos basin in North China and the Sichuan basin in Southeast China.The geothermal metamorphism of coal occurred universally in China, but both the coal measures and their overlying strata are mostly rather thin, therefore most coals that underwent only geothermal metamorphism are of low rank.

The geothermal state in coal basin is always related to the geolocial texture and structure of the Earth's crust[2-4,6]. The present pattern of isogeothermal contours in the Sichuan basin are similar to those of the shape of basin. Towards the central part of the Sichuan basin, the Moho depth in Nanchong and Neijiang is less than 38 km, shallower than the surrounding areas by about 2 km. The distribution of coal rank within the basin is obviously concordant with the Moho depth[17].

In the Ordos basin, the previously formed structural frame was reworked and adjusted by the Indosinian-Yanshanian movement. Simulation of dynamic geological-thermal history shows that the geothermal field of the basin influenced the geothermal metamorphic coals inhomogeneously. In this huge Mesozoic basin, the geothermal flux increased from the margin to the basin center, and the fitting data of the paleoheat flow changed approximately from 66.85 mW/m^2 to 41.80 mW/m^2 from Cambrian to the present (Fig.1); while in the southern part of the basin, the paleoheat flow conspicuously is enhanced to 91.12-99.9 mW/m^2 with the increasing subsiding amplitude and the coal ranks are also raised(Fig.2). Obviously the Triassic geotemperature was higher than pre-Triassic and post-Triassic ones. Therefore the spatial and chronological variation in the inhomogeneous paleogeothermal field together with the amplitude of subsidence controlled the evolution of geothermal coal metamorphism.

The paleogeothermal state in East China is different from that of Middle-West China. Owing to plate tectonics, the Earth's crust in East China has been stretched and thinned and the paleogeothermal flux strengthened, while in West-Middle China, especially in West China, the

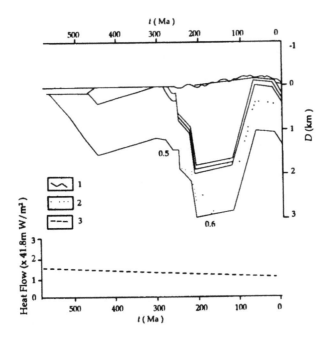

Fig. 1 Coalification pattern in the NE Ordos basin. 1, curve of subsidence; 2, $R_{o,max}$(%); 3, change of heatflow.

paleogeothermal flux is weakening due to the thickening of Earth's crust under tectonic compression. Besides, the intense tectonic uplift caused a shallow burial depth for coal-bearing strata and led to geothermal metamorphism of coal to reach a "standstill" state earlier in East China. On the contrary, the coals in the mega-superimposed basin in Central West China were being subject to geothermal metamorphism under deeper burial, hence higher ranks developed.

Telemagmatic Metamorphism of Coal
Telemagmatic metamorphism, another significant coal metamorphic type in China, can be subdivided into three subtypes, i.e. hypabyssal, mesogenetic and plutonic telemagmatic metamorphism. It is the Mesozoic and Cenozoic magmatic activities, especially igneous intrusions during the Yanshanian stage that had exerted important influence on part of geothermal metamorphic coal in China, so it is the main cause that produced large quantities of medium-high rank coals in China. The Yanshanian magmatic activities are characterized by their great intensity, wide sphere of influence and polycyclicity.

Fault belts control the magmatic intrusive. In most cases, magma follows a fault belt to intrude upward. In addition to physical data, the cutting depth of fault can be deduced according to the granite origin. Extensive occurrence of remelted granite is a symbol of the fault cutting into the sial; occurrence of mantle derived granite is a symbol of the fault cutting into the lithosphere; and widespread synthetic granite indicating the fault cutting into the sima. Especially in East China, the control of fault belt on magmatic intrusive is more obvious. For example, a series of fault

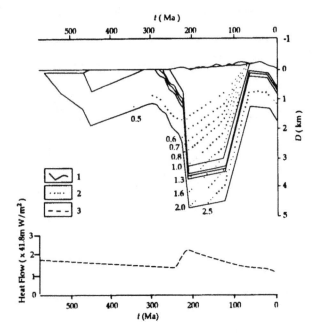

Fig. 2 Coalification pattern in the SE Ordos basin, 1, curve of subsidence; 2, $R_{o,max}(\%)$; 3, change of heatflow.

belts, such as the Sihui-Wuchuan, Lianping-Enping, Heyuan-Dongguan, Fogang-Fengliang and Gaoyao-Huilai develop in central-south Guangdong province, and the widespread Yanshanian adamellite and biotite granite are distributed adjacent to these fault belts, where intensive magmatic activities highly raised the metamorphic degree of Permo-Carboniferous and Triassic coals in Yingwong, Guanghua-Gaoyao,Yangehun,Taikai and Enping coalfield to anthracite through superimposed telemagmatic metamorphism(Fig.3).

In China, Mesozoic magmatic activities, especially the Yanshanian intrusions accompanied by telemagmatic metamorphism occur intermittantly along latitudinal and NE-NNE directions which are mainly related closely to fault belts as mentioned above. Most medium-high rank coals in China are distributed along these two directions and their intersections.

Medium-high rank coals were caused by telemagmatic metamorphism distributed intermittantly along the latitudinal direction. So far recognized are mainly as follows:1) along 40°30′ -42°30′ N. lat.; 2) along 40° N.lat.; the above two latitudinally distributed medium-high rank coals nearly pass through intermittantly from west to east China. 3) along 37° -38° N.lat.; 4) along the Huayinshan fault belt; 5)along the Longmenshan fault belt; and 6) along 24° -26° N.lat.

Medium-high rank coals were caused by telemagmatic metamorphism along NE-NNE directions. So far recognized are listed mainly as follows. In Northeast China they are: 1) along the Yilan-Yitong fault belt; 2) along the Mishan-Dunhua fault belt.In North China they are: 3) along east foothill of the Taihangshan fault belt; 4) along the Tancheng-Lujiang fault belt. In South China they are: 5) along the Huayinshan fault belt; 6)along the Longmenshan fault belt; 7) the Gongtian-

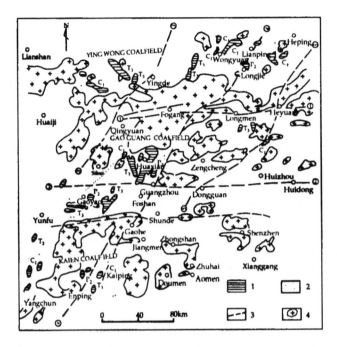

Fig.3 Diagram showing the relationship between coal metamorphism and tectonic activities in central-south Guangdong province. 1,anthracite; 2,coalfield boundary; 3,fault belt; 4,intrusive; ①,Sihui-Wuchuan fault belt; ②,Lianping-Enping fault belt; ③,Heyuan-Dongguan fault belt; ④,Fogang-Fengliang fault belt; ⑤,Gaoyao-Huilai fault belt.

Ningxian-Xinning fault belt; 8) the Lingyunshan-Hengyong-Yongzhou fault belt; 9) the Dongxian-Yichun fault belt. In Southeast China they are: 10) the Xiaoshan-Qiuchuan fault belt; 11) the Jianou-Yongan fault belt; and 12) the Lianping-Enping fault belt.

The Yanshanian magmatic activities are more intense in the south and east provinces than in the north and west, consequently most intense in SE China, i.e. the strongest coal metamorphism occurs in SE China.

Hydrothermal Metamorphism of Coal
Hydrothermal metamorphism caused by differential magmatic hydrothermal solution is often associated with telemagmatic metamorphism, as the high rank coals occur in west Henan province and Wumuchang mining district in Inner Mongolia Autonomous Region. Moreover, there is another type of hydrothermal metamorphism of coal found in China. The heatsource of metamorphism is confined water under high temperature from the crustal depth instead of differentiated magmatic solution. A typical example is found in the south part of Reshui coalfield of Qinghai province, where three middle Jurassic coal seams are counted upward No.1 to No.3, and no post-coal-forming igneous rocks have been found in the coalfield and adjacent region. The first characteristic of such a hydrothermal metamorphism is that the coal rank of the middle seam(No.2) may reach as high as 2.23% $R_{o,m}$, higher than both the seam No.1 and the overlying

No.3. Such an anomaly obivously runs counter to Hilt's Rule. Another feature is that the coal rank of seam No.2 tends to change with the flowing direction of subsurface hot water and the distance of channel way, thus resulting in that the coal rank of seam No.2 varies both along strike and dip.

The mechanism of this type of hydrothermal metamorphism is that the Datong mountain to the north of Reshui coalfield supplies water into the great depth of the crust where the water heated to 400-500 °C, and then asending as high-temperature confined water along the fault F_0 situated to the south of the Reshui coalfield. The ult F_0 is the first grade channel way for the hot confined water; the two set of faults cut obliquely fault F_0, the coal seam roof, and the cleats plus micropores within the coal are the second, third and fourth channel ways respectively. The roof of seam No.2 is a permeable coarse sandstone which serves as good channel way of the third grade. That is why the coal rank of seam No.2 is higher than that of seam No.1 and No.3, the roof of the other two seams being impermeable mudstone. Figure 4 shows the distribution of hot springs in Reshui coalfield, and figure 5 shows roughly the circulation of hydrothermal fluid in the Datongshan-Reshui coalfield. Perhaps this kind of hydrothermal metamorphism may better be called "hot water metamorphism of coal", so may differentiate from the other type of hydrothermal metamorphism of coal. As the contact metamorphism influences the coal only locally in China, it is left out here.

MULTISTAGE METAMORPHIC EVOLUTION AND SUPERIMPOSED METAMORPHISM THROUGH MULTITHERMO-SOURCES IN CHINESE COALS

The main cause for the formation of coal-rank distribution with various ranks is that Chinese coals generally underwent multistage evolution and multithermo-sources superimposed metamorphism.

Multistage Metamorphic Evolution
Three evolutionary stages with their main features are as follows.

The first evolutionary stage is characterized by the predominance of geothermal metamorphism of coal. Chinese coals universally underwent geothermal metamorphism diagenesis, but most Tertiary coals remain in the diagenetic stage.The ranks of geothermal metamorphic coals are controlled by geotemperature and its duration. The geotemperature under which coal seam has suffered depends on the subsiding depth of the seam as reflected by the thickness of the overlying strata. The Permo-Carboniferous coal measures in North China is widely distributed, and its thickness is as a whole rather thin with little variation except in a few areas. The Triassic system was so universally developed in North China that the geothermal metamorphic degree of Late Paleozoic coals in China depended mainly on the thickness of the overlying Triassic (Fig.6). For example, the cover of Late Paleozoic Shanxi Formation in central and western Shandong province is about 1000-1500m thick, about 1700-2100m thick in the Handan-Xingtai coalfield, less than 1500m thick in the Huaibei and Huainan coalfields,about 1500m thick in the Pingdingshan coalfield. A large amount of Late Paleozoic coals not influenced by abnormal heat in North China generally hasn't exceeded the fat coal stage(Table 1), except that in a few areas, such as in the Houma basin, Shanxi(4272m thick), and in Jiyuan (4499m), Henan, where much thicker coal measures and overlying strata developed, relatively high rank coal occurs. The thickness-

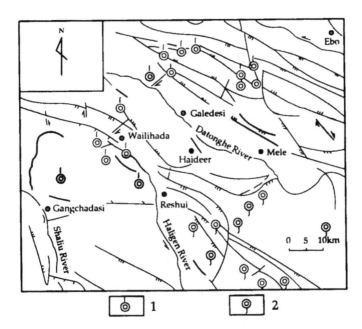

difference of coal measures and overlying strata resulted in the development of different coal ranks. For instance, in Mianzhu, Sichuan province owing to the uplifting of the Longmenshan during the Indosinian movement, the total thickness of the Permian Longtan coal measures and its overlying strata is about 1200m, and coal ranks only reached the gas to gas-fat coal stage; whereas near the tectonically undisturbed Chongqing situated southeast to Mianzhu, the corresponding thickness reached about 4650m, and the rank evolved to the coking coal stage. In another example, the Late Jurassic-Early Cretaceous coal measures of Zalainuoer,Yimin and Helinghe coalfields of east Inner Mongolia Autonomous Region, is about 1000-1400m thick, and the coal remains in the lignite stage, while in the Hegang, Boli and Jixi coalfields of Heilongjiang province, the coeval coal measures is about 1500-2300m thick,reaching a total of 3000-4000m with addition of the Cretaceous thickness, and the coal evolved to gas coal stage. Besides, the later magmatic intrusion changed a part of the low rank coals into the medium-high coal rank stage. The larger the amplitude of subsidence,the higher coal rank is, which is typical in the multistage coalfield; furthermore, the effect of time is well displayed in Shiguaigou of Daqing mountain, Inner Mongolia Autonomous Region, where the Late Carboniferous coals reach fat to gas coal stage; the Early-Middle Jurassic coals reach gas coal stage, while the Early Cretaceous coal still remains as lignite.

The second evolutionary stage is characterized by superimposed metamorphism through multithermo-sources. It is generally known that low rank coals were mainly formed at the first evolutionary stage. In China there are quantities of bituminous and anthracite among which normal geotemperatures generally do not differ as much as to cause big disparity in the degree of coal metamorphism. On the other hand, in the Laiwu coalfield of Shandong province, the

Yang Qi et al.

geotemperature at the depth of 300m above a Yanshanian intrusive is 3-5°C higher than that of neighboring areas, and the shape of geothermal contours is concordant with the outline of the Yanshanian granodioritic body lying at 400-800m below the ground surface, where the coal metamorphic zone of medium-high coal rank occurs. The Lianshao coalfield of the Permian Long-

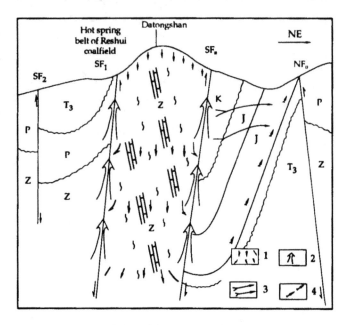

Fig.5 Sketch map showing the circulation of hydrothermal fluid in the Datongshan-Reshui coalfield. 1,recharge water; 2,pathway of first order; 3, parthway of third order; 4,pathway of fourth order.

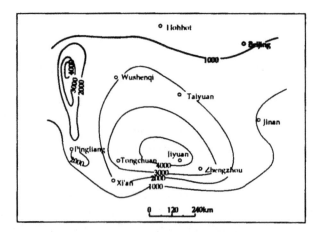

Fig.6 Thickness contour of the Permian-Triassic strata in North China(unit:m).

Table 1 The ranks reached by geothermal metamorphic coals in some coal mines, North China

Area	Geological Age	$R_{o,max}$(%)	Coal Rank
Hequ	P_1	0 65	long flame coal
Zhaogezhuang	P_1	0 79	gas coal
Yanzhou	P_1	0.71	gas coal
Hunyuan	P_1	0 58-0.60	long flame coal
Donghe	P_1	0 79	gas coal
Feicheng	P_1	0 77	gas coal
Xinwen	P_1	0 78	gas coal
Xuzhou	P_1	0 77	gas coal
Xuchang	P_1	0 65	long flame coal
Hebi	P_1	0.72	gas coal

tan Formation in central Hunan province covers the north and south synclines. Around it Indosinian and Yangshanian intrusives are discretely distributed, and also along 27°30′ N.lat. between the two synclines, so that both the north and south synclines of the coalfield are encircled by discretely distributed Mesozoic intrusives, which caused the coal ranks to decrease towards the center in both syncline. The above two examples indicate that in the second evolutionary stage it is mainly the superimposed telemagmatic metamoprhism that promoted part of low rank coals formed during the first evolutionary stage to medium-high rank ones. Besides the large-scale magmatic intrusion, the other heat sources that might have raised coal rank during the second evolutionary stage, are radioactive heat from intrusives, and high temperature conducted upward through deep fault.

The abnormal geothermal field resulting from the uplifting of Moho depth should be mentioned. The area with shallower Moho depth usually is accompanied with medium-high rank coals. The crustal thickness usually changes abruptly along the belt with distinct difference in Moho depth, such belt is often the boundary between tectonic units with long-term developing faults combined with rather intensive magmatic intrusion, and the coal nearby is raised to a higher rank. For instance, the medium-high rank coals occur in Hanxing and Hebei coalfields near the gravity gradient zone of the Taihang ranges; the high rank coals in Hulusitai, Ruqigou and Zhongning coalfields near the gravity gradient zone of Yingchuan-Liupanshan.

The third evolutionary stage is characterized by the establishment of coal metamorphic frame. Raising the coal rank is no more important in the third evolutionary stage than such features as the displacement of coal metamorphic zonation already formed by the Cenozoic tectonic movement, and the establishment of coal metamorphic frame in China. It is the NE-NNE deep faults that remarkably reworked the coal metamorphism zonation already formed; the most typical case is the dextral shifting of the Zijingguan fault (Shanxi-Hebei provinces) trending NNE, which removed the high rank coal zones in the Taihang ranges and central Shandong province to the east of the fault southward by about 1° latitude relative to the northern end of the Qingshui basin to the west of the fault; the shifting distance is about 100km from 38°-37° N.lat.to 37° -36° N.lat.The high metamorphic coal zone in central Henan province joined originally with high metamorphic coal zone in southeastern Shanxi province to the west of the Zijingguan fault zone was also removed southward(Fig.7). The clockwise shearing of the above faults trending NE-NNE is owing to the

NNW-ward   subduction of the Pacific plate[8]. The influence of the Pacific plate activities on the
Tancheng-Lujiang fault belt is more obvious because the fault located closer to the east. The
Tancheng-Lujiang fault zone was compressed from east and west accompanied with a dextral
shifting since the Neogene[7,18]. Other similar NNE-NE deep fault zones are the Huangyan-
Huilai fault zone, the Zhenjiang-Haifeng fault zone, the Yuyao-Shenzhen fault zone, the
Lianping-Enping fault zone and the Sihui-Wuchuan fault zone. There are high rank coals
metamorphosed through magmatic intrusions in the vicinities along these fault zones. The Early
Tertiary sinistral shifting of the Zhejiang-Haifeng fault zone in southeast China is also verified[7].

Superimposed Metamorphism through Multithermo-Sources in Chinese Coals

Coal in the same coalfield may undergo one or more other types of coal metamorphism
superimposed on the basal geothermal metamorphism(and diagenesis), and may also suffer the
same type of metamorphism more than once. For these cases it is suggested to use the expression
"superimposed metamorphism through multithermo-sources of coal". So far five types of
superimposed coal metamorphism recognized in China are as follows.

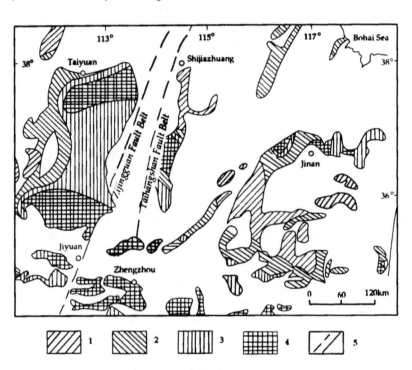

Fig.7 Map showing the dextral shifting of Zijingguan fault belt removed the east side rank zonation
southward.1,low rank bituminous; 2,medium rank bituminous; 3,high rank bituminous; 4,anthracite; 5,fault
belt.

Type I is geothermal metamorphism superimposed by telemagmatic metamorphism. A good
example is the 130m thick Paleocene coal of Fushun coalfield, where this Tertiary coal has
already been coalified to the long flame-gas coal stage. Through mathematical simulation, the

thermal gradient ever reached in the Fushun coalfield may   enhance the rank of extrathick coal seam reaching 0.40% $R_{o,m}$, still remain in brown coal stage. It is the Oligocene diabasic magma intruded into the underlain   stratum that promoted the rank of the ultrathick coal seam to 0.6%-0.75% $R_{o,m}$, and the simulation coincides with the actually measured data.

Type II is geothermal metamorphism superimposed more than once by telemagmatic metamorphism. The Datian coal district in Fujian province was subject to the influence of magmatic activities at least twice during the Indosinian-Yanshanian stage after the deposition of the Permian coal measures, thus the coal rank has been raised to 5.64% $R_{o,m}$.

Type III is geothermal metamorphism superimposed by telemagmatic and contact metamorphism. Differing from superimposition simply by telemagmatic metamorphism, a part of the coal seam was intruded by magma and the coal changed into semigraphite or graphite under well-confined condition or changed into natural coke under badly confined condition, besides the thermal effect by telemagmatic metamorphism.

Type IV is geothermal metamorphism superimposed by hydrothermal metamorphism. A typical example is the metamorphism of the Reshui coalfield, where no magmatic activity occurred after the coal-forming period, the coal in the north district of the coalfield which experienced only geothermal metamorphism reached to the long-flame coal stage; while in the south area, the bordering deep fault served as the channel way for the hot confined water which is the thermo-sources responsible for raising vitrinite reflectance of the coal to 0.72%-2.23% $R_{o,m}$.

Type V is geothermal metamorphism superimposed by telemagmatic and hydrothermal metamorphism. The coal in the Jiafu coal mine ranking as 1.8%-2.0% $R_{o,m}$ underwent only geothermal metamorphism, and mathematic simulation indicates that magmatic intrusion raised the coal rank to 4.8%-5.0% $R_{o,m}$, it is the superimposed hydrothermal metamorphism which finally raised it to 5.44% $R_{o,m}$.

Multievolutionary stages and superimposed metamorphism through multithermo-sources are the most prominent features of Chinese coal metamorphism. Geothermal metamorphism together with various kinds of superimposed metamorphism brought about various kinds of coals which fulfill the numerous usage for energy resources in China.

THERMAL DYNAMICS ANALYSIS OF COAL METAMORPHISM IN CHINA

According to the   general principles of geothermics, the characteristics of overall geothermal state that controlled the evolution of coal and dispersed organic matter can be described by using paleogeothermal flux value and direction of each spots in coal-bearing and oil-gas-bearing strata[1,9-11]. The paleogeothermal flux of coal-bearing and oil-gas-bearing strata is the function of specific paleothermal conductivity and paleogeothermal gradient. The strata paleogeotemperature is the main controlling factor of coalification degree, and paleogeothermal gradient is the main controlling factor of coalification gradient, therefore all factors that influenced paleogeotemperature and its gradient can be regarded as factors that affected the degree and gradient of coalification. This is the theoretical basis for thermal dynamics analysis of coal metamorphism.

From the point of view of equilibrium, and disruption of geothermal state, the geothermal flux and additive abnormal thermal flux are synthesized, while two-dimension dynamics models of geothermal coal metamorphism and magmatic coal metamorphism based on the history of both subsidence and thermal evolution are set up through theoretical inference. In addition, by applying the method of paleogeothermal structural analysis of earth crust, an empirical formula T-t-$R_{o,m}$ of thermal dynamics of coal metamorphism is put forward[10,11]. The formula T-t-$R_{o,m}$ is $T=e^{[646.32/(\ln t+111.85)]-(0.492t 0.093/R_{o,m})}$ where, T is the paleotemperature (°C); t, absolute age of stratum; $R_{o,m}$, mean vitrinite reflectance(%).

The Fushun basin and the east part of Zhejiang-Fujian-Guangdong provinces are quantitatively interpreted as examples of superimposed metamorphism,and a mathematic simulation of dynamic state of multistage evolution and superimposed coal metamorphism through multithermo-sources is preliminarily accomplished.

Fushun Basin
The coal seam of gigantic thickness occurring in the Paleocene Guchengzi Formation in the Fushun basin reached the flame coal-gas coal stage, which is much higher than other coeval coal seams. Through two-dimensional dynamic state modelling of coal metamorphism in the Fushun basin, the paleogeothermal structure during the formation of Fushun basin is determined(Table 2), and on this basis the geothermal gradient of the Fushun basin is also calculated, under the influence of this geothermal gradient, up to now the coalification of the extrathick coal seams may reach at most the lignite stage($R_{o,m}$=0.40%). In addition, the modelling results prove that the super-coalification in the Fushun basin is due to the abnormal geotherm provoked by the large-scale Late Oligocene diabasic intrusion after coal accumulation, thus raising the coal rank of the Guchengzi Formation to the flame-gas coal stage ($R_{o,m}$=0.5%-0.80%)(Fig.8,9).

Table 2. Analysis data and results of paleo-geothermal texture of the Fushun basin

Horizon		H(m)	K(W/m. °C)	A(W/m²)	Q(mW/m²)	T(°C)
surface					82.0	10.2
	E_2	0 5	2.7	0.8	81.8	25.3
upper crust	E_1	1 1	1 45	1.8	81.3	58.7
	K	1 6	2 1	1.05	81.0	78
	Ar	10.0	2.2	0.73	78.0	364
middle crust	granite	16 0	3 1	2 45	70 6	487
lower crust	mafic rock	31.8	2.5	0.31	68.0	927

H,depth of lower boundary;K,specific thermal conductivity of radioactive elements;Q,thermal fluid value oundary;T,temperature at lower boundary

form belt of high-rank coal,   such as the anthracite zones in the southern margin of North China

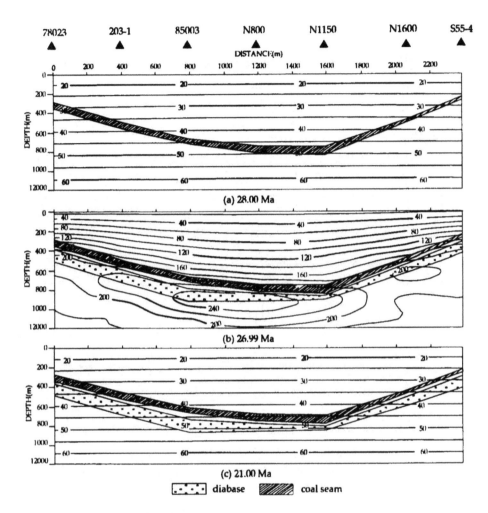

(a) 28.00 Ma

(b) 26.99 Ma

(c) 21.00 Ma

diabase   coal seam

Fig.8 Geotemperature distribution of the bottom of the Guchengzi Formation along the exploration line E1600 under the effect of diabasic sill in the Fushun basin.

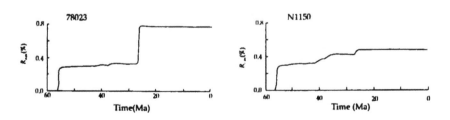

Fig.9 Vitrinite reflectance variation curve of the bottom of the Guchengzi Formation along the exploration line E1600 under the effect of the diabasic sill in the Fushun basin.

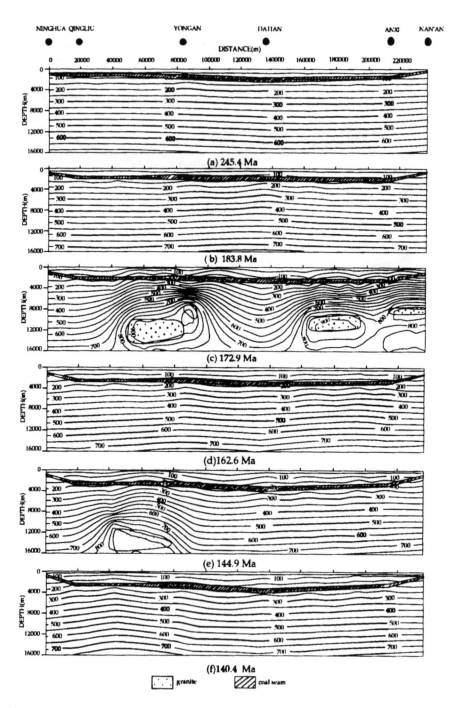

Fig.10 Geotemperature distribution along the Quanzhou-Ninghua section before and after the granite intrusions.

Southeastern Coastal Provinces

By the same method, we have also modelled:the rising temperature and coalification of Upper paleozoic coal seams under normal subsiding conditions; the heat conduction and temperature lowering process after granitic magma intruded. The rising temperature and metamorphism of coal under large-scale magmatic intrusives in the eastern parts of Zhejiang-Fujian-Guangdong provinces shown in Quangzhou-Ninghua section(Fig.10). Through modelling, the paleogeothermal structure of earth crust around superimposed basins is also calculated, testifying that the coal rank in the Lower Permian Tongzhiyan Formation under the geothermal metamorphism can only reach fat coal to low-rank anthracite stage, the vitrinite reflectance may reach as high as 3.27%. But under the much higher temperature provoked by later multistage large-scale granitic intrusions, the coal seam could reach medium-rank anthracite stage as at Anxi, and to meta-anthracite stage as at Yong'an(Fig.11). Obviously, the abnormal increase of coal ranks in the study area was caused by the strong multi-stage telemagmatic metamorphism superimposed on geothermal metamorphism of coal.

DISTRIBUTION REGULARITIES OF COAL METAMORPHISM IN CHINA

The distribution frame of coal metamorphism in China is inseparable from the formation and evolution of the tectonic frame, especially the crucial influence of the Indosinian, Yanshanian and Himalayan movements.

The Coal Metamorphism Displays Obvious Differences betwwen North and South, East and West in China

The metamorphic degree of coal tends to increase from north to south. Geothermal metamorphism dominates the northern part of China, and the superimposed metamorphism through anomaly-heat increased in intensity and extension towards south, especially in southeast China.

China may be divided into west, middle and east parts with Daxing'anling-Taihangshan-Wulingshan as the boundary between west and middle parts, and the Tancheng-Lujiang deep fault belt as the boundary between middle and east parts. The superimposed metamorphism provoked by anomaly-heat increases both intensively and extensively from west to middle and to east, and this is in accordance with the thinning of the Earth crust, shallowing of Curie isotherm plane and strengthening of tectonic-magmatic activity from west to east. It is the result of multiple subducting of the Pacific plate towards the east Eurasian continent and the influence is weakening westwards.

The Margin of the Subarea is Characterized by Superimposed Metamorphism through Multithermo-Sources

The boundary of coal metamorphic subarea usually is the convergent margin of plates where block activity, magmatic activity and changing of crust's thickness are most intense. It leads to

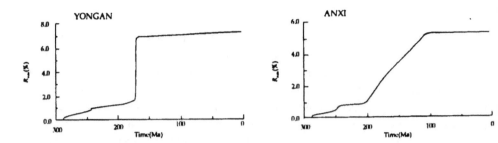

Fig.11 Vitrinite reflectance variation curve of the roof of the Lower Permian coal measures affected by granite intrusion along the Quanzhou-Ninghua section.

the high-rank anthracite zone in southeastern China.

Block Structure is a Decisive Factor Controlling Zonation of Coal Metamorphism inside the Subarea of Coal Metamorphism

The coal rank within fault block is relatively low, but it is higher along the fault zone, especially around the multistage-activated deep fault belt which cut the Moho discontinuity. The Tertiary coal neighbouring the Yarlungzangbojiang deep fracture in Tibet, has already been metamorphosed into anthracite.

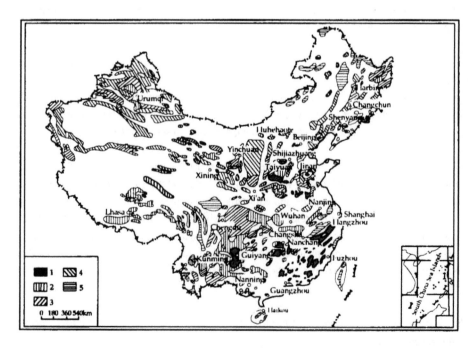

Fig.12 Distribution of coal ranks during the main coal-forming periods in China. 1,anthracite;2,high metamorphic bituminous; 3,medium metamorphic bituminous; 4,low metamorphic bituminous; 5,lignite.

All Late Paleozoic coals in China have exceeded the diagenetic stage and medium-high rank coals reach an important percentage mainly due to superimposed metamorphism. Low rank bituminous coals predominate in the Mesozoic, but lignites which not yet reach metamorphic stage are also present. Most Tertiary coals remain as lignites in the diagenetic stage, except a few local occurrence of low-rank bituminous. The spatial distribution of coal ranks in China shows a general tendency of increasing from north to south and from west to east. Multithermo-sources superimposed metamorphism and especially the superimposed telemagmatic metamorphism since the Yanshanian orogeny have evidently reformed the Chinese coal ranks; the Tertiary tectonic displacement brought about the distribution frame of the Chinese coal-rank zonation(Fig.12).

In a word, the distribution of coal metamorphic zone in China is controlled by tectonism and it reflects comprehensively superimposed metamorphism through multithermo-sources and multistage evolution. Fig.12 shows the distribution of coal ranks during the main coal-forming periods in China.

REFERENCES

1. Bostick,N.H.Thermal alteration of clastic organic particles as an indicator of contact and burial metamorphism in sedimentary rocks, *Geosci.Man.***3**,83(1971).
2. Cochran,J.R.Effects of finite rifting times on the development of sedimentary basins, *Earth Plant.Sci.Lett.***66**, 289-302(1983).
3. Falvey,D.A. and Middleton,M.F.Passive continental margins:evidence of a prebreakup deep crustal metamorphic subsidence mechanism. In:Colloquium on Geology of Continental Margins(C₃),*Oceanologica Acta* **4**, 103-114(1981).
4. Mckenzie,D.Some remarks on the development of the sedimentary basins,*Earth Plant.Sci.Lett.***40**, 25-32 (1978).
5. Pan Weier,Yang Qi and Pan Zhigui. Magmatic thermo-metamorphism of coal in central-southern Hunan and Jianxi provinces, *J.Geoscience* **7**, 326-336(1993).
6. Sawyer, D.S. Effects of basement topography on subsidence history analysis, *Earth plant.Sci.Lett.* **78**, 427-434(1986).
7. Wan Tianfeng.*The stress field of deformation structure and its application in Mesozoic-Cenozoic plates of east China*, Geological Publishing House, Beijing(1993).
8. Wang Hongzhen, Yang Shennan and Liu Benpei(Eds). *Tectonic palegeography and biologic paleogeography in China and adjacent areas*, Press of China University of Geosciences, Wuhan(1990).
9. Wang Jiyang. Thermal texture of crust-upper mantle of the Liaohe rift basin, *Science in China Series B* **29**(1986).
10. Wu Chonglong, Li Sitian and Cheng Shoutian.The statistical prediction of the vitrinite reflectance and study of ancient geothermal field in Songliao basin, China, *J.China Univ.Of Geosci.***2**, 91-101(1991).
11. Wu Chonglong, Zhou Jianyu and Wang Gengfa. Tectonic and sedimentary history of southeastern coast region, China:a synthesis, J.*China Univ.of Geosci.***6**,64-79(1995).
12. Yang Qi. Coalification. In:*Advances in Coal Geol.*, Yang Qi(Eds). Science Press, Beijing(1987).
13. Yang Qi. Study on coal metamorphism in China,*J.China Univ. of Geosci.***14**,341-345(1989).
14. Yang Qi. Study on coal metamorphism, *J. Geoscience*, **6**, 437-443(1992).
15. Yang Qi, Pan Zhigui, Weng Chengmin, Su Yuchun and Wang Zhenping. *Metamorphic characteristics and geological causes of permo-Carboniferous coal in North China*, Geological Publishing House, Beijing(1988).
16. Yang Qi and Ren Deyi. Basic characteristics of coal metamorphism in China, *Coal Geology and Exploration.* **1**,1-10(1979).

17. Yang Qi, Wu Chonglong, Tang Dazhen, Kang Xidong and Liu Dameng. *The coal metamorphism in China*, Coal Industry Publishing House, Beijing(1996).
18. Zhang Wenyou, Ye Hong and Zhong Jiayou. Fault blocks and plates, *Science in China Series B* 21(1978).

Proc. 30th Int'l Geol. Congr., Vol. 18, Part B, pp. 77-98
Wang Qi (Ed.)
© VSP 1997

Variations in Coal Rank Parameters with Depth Correlated with Variscan Compressional Deformation in the South Wales Coalfield.

ROD GAYER and RICHARD FOWLER

Laboratory for Strain Analysis, Department of Earth Sciences, University of Wales Cardiff, PO Box 914, Cardiff, CF1 3YE, UK.

Abstract

Coal maturity data in the form of volatile matter (%Vm_{daf} and %Vm_{dmmf}) and random vitrinite reflectance (%Ro) have been analysed for the South Wales coalfield. They show that in general coals increase in rank with depth, obeying Hilt's law, and increase in rank laterally from high volatile bituminous coals in the south and east of the coalfield to anthracite in the north-west of the coalfield. Coal rank was acquired both before and during Variscan deformation of the coal-bearing foreland basin. The rank pattern with depth in the eastern half of the coalfield suggests a palaeogeothermal gradient of approximately 218 °C km^{-1}. Detailed analysis of both %Vm_{daf} and %Ro data from individual collieries reveals the presence of excursions from Hilt's law which affect one or more coal seams within the Westphalian A - lower Westphalian C sequence. It is shown that the excursions commonly correlate with thrust detachments within the coal seams. Detailed petrological and structural analysis of a sequence of coal seams from Llanilid West Opencast coal site show a complex pattern of variation in %Vm_{daf} and %Ro, similar to the maturity variations recorded in boreholes through the Westphalian Coal Measures in the Netherlands. However, in South Wales no correlation with coal composition is apparent, but a strong correlation with a newly defined Numerical Deformation Index (NDI) suggests a relationship between coal maturity parameters and deformation. The high palaeo-geothermal gradients present in the coal-bearing sequence have been modelled using BasinMod* software. Initial results suggest that a basal heat flow of 63 mWm^{-2}, normal for a continental crust setting, was enhanced by a lateral heat flow of 33 mWm^{-2} from hot fluids passing laterally through the basin. It is argued that the excursions in rank maturity from Hilt's Law represent an increase in temperature caused by fluids carrying heat into the coal seam along seismically active thrusts.

Keywords: coal rank, vitrinite reflectance, thrusting, maturity modelling, Westphalian

INTRODUCTION

The increase in coal rank with depth (Hilt's Law) has been confirmed by numerous studies since the relationship was proposed and is a consequence of the early stages of burial metamorphism. The predominant factor appears to be temperature [32], although various studies suggest that both pressure [22] and time [33] can also influence the process. The two commonly used coal rank parameters are volatile matter (dry and ash free - %Vm_{daf} or dry and mineral matter free - %Vm_{dmmf}) and vitrinite reflectance (random in oil - %Ro). Abnormal variations in %Ro with depth in boreholes [10] and in individual coal seam profiles [35] have been documented in the Netherlands; in both cases attributed to original redox conditions in the precursor peat [34]. Although M. Teichmüller & R. Teichmüller [32] indicated anomalous %Ro associated with minor thrusts in the Ruhr coal district, relatively few subsequent studies have investigated the relationship between thrusting and coal rank development [8, 39]. Recently, Gayer et al. [17] described %Vm_{daf} excursions from Hilt's Law in 94 out of 154 (61 %) colliery data sets from the South Wales coalfield. Several of these excursions were correlated with thrust detachments in the

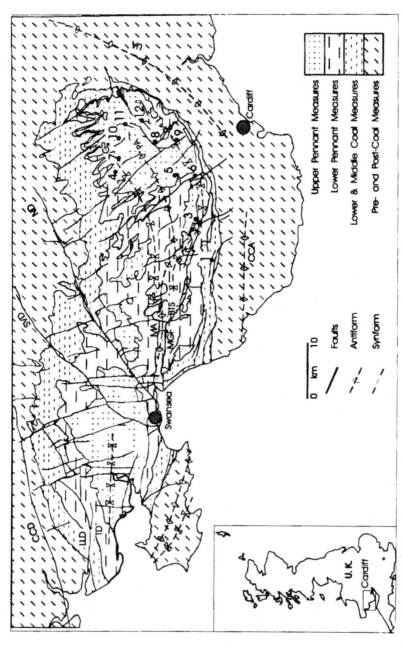

Figure 1. Location map of the South Wales Coalfield.
CCD Carreg Cennan disturbance; LLD Llannon disturbance; ND Neath disturbance; SVD Swansea Valley disturbance; TD Trimsaron disturbance; MGF Moel
Gilau fault; BTS Bettws Tonyrefail syncline; CCA Cardiff Cowbridge anticline; GS Gelligaer syncline; LCS Llantwit Caerphilly syncline; MA Maesteg anticline
PA Pontypridd anticline; UA Usk anticline. Location numbers are colliery sites except 1 which is an opencast site. 1 Llanilid West; 2 Coedely; 3 Cwm No 4 shaft;
4 Groesfaen; 5 Windsor; 6 Nantgarw; 7 Bargoed; 8 Wyllie; 9 Bedwas; 10 Oakdale; 11 Celynen North; 12 Celynen South. Inset shows coalfield location within the

coals or associated mudrocks, and tentatively ascribed them to localised heating by hydrothermal fluid migration along the thrusts.

This paper presents new data from collieries and an opencast coal mine in the south-east of the South Wales coalfield, documenting variations in %Vm$_{daf}$ and %Ro with depth and %Ro through individual coal seams. It also presents the results of coal maceral analyses from the same coal seams and introduces a numerical deformation index (NDI) which attempts to quantify the development of compressional deformation fabric within the coals. The possible relationship between coal rank development, thrusting, and basinal fluid flow is also discussed.

REGIONAL GEOLOGY OF THE SOUTH WALES COALFIELD

Stratigraphy
The South Wales coalfield forms an erosional remnant of a major Late Carboniferous coal-bearing foreland basin [14, 26]. The sediments in the basin are preserved in a structurally complex E-W trending Variscan synclinorium extending from SW Dyfed to the western flank of the Usk antiform (Fig. 1). The coal basin overlies a southward thickening (0 - 1 km) Lower Carboniferous platform carbonate sequence [38] which passes conformably downwards into a thick (3 km) Old Red Sandstone unit and a shallow marine Lower Palaeozoic succession. Crystalline basement underlies the coal basin at depths ranging from 3.5 km in the north-west of the main coalfield to over 6 km in the east [21]. The Coal Measures sequence is up to 3.5 km thick in the centre of the basin and ranges in age from basal Namurian to early Stephanian. The basin was initiated during the early Namurian, following a regional compressional event that resulted in the break-up of the Dinantian carbonate platform, the relocation of the basin depocentre and the influx of clastic detritus [18]. Throughout Silesian sedimentation, the basin depocentre was oriented approx. E-W (varying from NE-SW to NW-SE) and centred on the Swansea - Gower area. Stratigraphical thicknesses decrease markedly away from the depocentre, particularly to the east but more gradually to the north and west [19]. The basin-fill sequence coarsens and shallows upwards from marine mudstones and sandstones (Namurian A - lower Westphalian A), through coastal plain coal-bearing mudstones and sandstones (upper Westphalian A - lower Westphalian C) to coarse grained sandstones and conglomerates deposited in an alluvial braidplain (upper Westphalian C - Stephanian) (Fig. 2) [19, 24]. The majority of the 125 coal seams present in the coalfield occur in the main productive Coal Measures of late Westphalian A - Westphalian B age.

Although no younger solid formations currently overlie the Upper Carboniferous Coal Measures, to the south of the coal basin, in the Vale of Glamorgan, a relatively thin (< 300 m) Mesozoic sequence (Upper Triassic - Lower Jurassic) rests unconformably on Variscan deformed Dinantian - Namurian rocks. Approximate vitrinite reflectance values of 0.5 %Rm. in the Lower Jurassic rocks suggest a possible younger Mesozoic cover subsequently eroded.

Structure
The basin-fill has been affected by Variscan deformation, the principal elements of which are: i) approximately E-W trending basin-scale north-verging folds; ii) strike-parallel thrusts (and associated lag faults), with a dominant northwards transport except along the southern margin of the coalfield where the thrusts are southward vergent; iii) ENE-WSW trending zones of fold and thrust disturbance, thought to represent reactivated basement Caledonoid structures as Variscan thrust ramps [7, 25]; and iv) NNW-SSE striking cross faults, commonly showing evidence of early strike-slip movement and later normal dip-slip movement (Fig. 1) [9]. All of these structures have been interpreted as the effects of north-north-west propagating Variscan fold and thrust deformation into the coal-bearing foreland basin [16]. Jones [23, 24] showed that sedimentation throughout the Westphalian was affected by the incipient development of the main E-W folds in

R. Gayer and R. Fowler

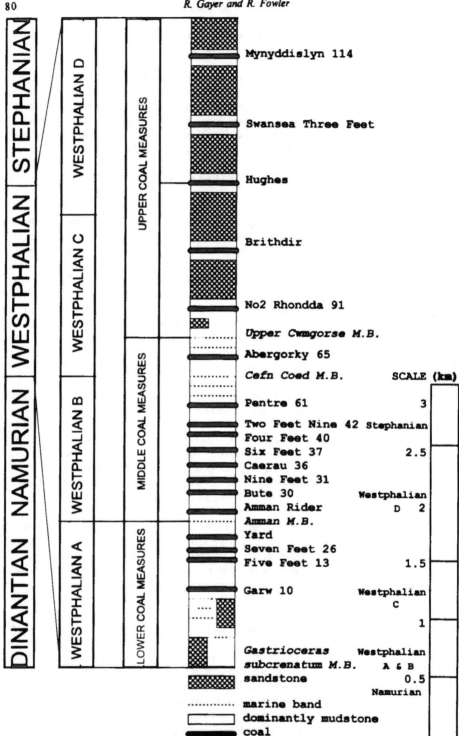

Figure 2. General South Wales Coalfield stratigraphy showing principal coal seams.
The inset shows the stratigraphic column to scale. Seam numbers after an unpublished report by R. Thewlis.

the coalfield, demonstrating the close timing between basin subsidence and compressive deformation.

Of particular relevance to this paper are the thrusts, which have been analysed in some detail within the working opencast coal mines [13]. This analysis has shown that thrusts occur either as major detachments along coal seams with imbricate thrusts branching upwards into the hangingwall sequence, or as isolated thrust ramps which in some cases are linked downwards to thrust detachments in the coal seams [20, 25]. Where thrust detachments interleave seat earth into a coal seam, an oblique fabric, locally termed rashings, is developed. Thrust detachments commonly occur within the coals of the lower part of the productive Coal Measures and it has been demonstrated that detachments within several coal seams have moved simultaneously producing a style of deformation that appears to be unique to coal-bearing sequences. This has been termed Progressive Easy Slip Thrusting (PEST) [12], and can be explained by the overpressuring of fluids generated as the coal seams are dewatered and matured by compaction, facilitating thrust propagation [15, 16].

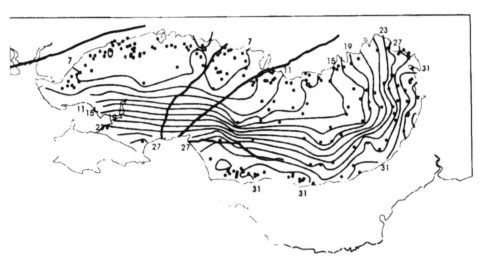

Figure 3. Coal rank %Vm$_{dmmf}$ isovols for the Four Feet seam, after White [37].

Metamorphism
The sediments of the coalfield have been affected by very low grade metamorphism within the diagenetic grades of metamorphism, based on illite crystallinity characteristics [36, 37]. The metamorphism ranges into the lower anchizone facies in the extreme north-west of the coalfield, based on the presence of pyrophyllite [6]. Coal rank, based on %Vm$_{dmmf}$ and %Ro, increase from high volatile bituminous coal in the south and east of the coalfield to anthracite in the north-west of the coalfield (Fig. 3) [36, 37], broadly coinciding with the illite crystallinity metamorphic pattern. White [36, 37] demonstrated that the coal isovols (lines of equal volatile matter) are parallel to stratigraphical boundaries around major fold structures in the coalfield, suggesting a pre-Variscan folding origin for the coal rank development, and Gayer *et al.* [17] showed that %Ro was developed both before and during thrust deformation. The rapid lateral increase in coal rank has been discussed in detail by White [36, 37] and by Austin & Burnett [2]. They suggested that the mechanisms in most agreement with the observations are: i) burial beneath a now eroded sedimentary load [37]; or ii) inflow of hot fluids either from the Variscan mountain belt to the south [16] or along deeply penetrating faults in the underlying basement [2].

VARIATIONS IN COAL RANK WITH DEPTH

Existing coal rank data for the South Wales coalfield consist mostly of measurements made by British Coal, both deep mines and opencast, or their predecessors, the National Coal Board, on the specific coal seams worked at each colliery. The vast majority of these recorded rank determinations from the coalfield are in the form of %Vm$_{daf}$ or %Vm$_{dmmf}$. This was the standard analysis carried out, commonly associated with a full proximate analysis. Within the last 10 years %Ro measurements were also occasionally made.

By combining the data from collieries lying in the same coal rank belt in the coalfield, a composite plot of rank versus depth below the No2 Rhondda coal seam (seam 91) has been produced for both %Vm$_{dmmf}$ and %Ro (Figs. 4a & b). Both plots show a similar increase in rank with depth, obeying Hilt's Law, but the data have considerable scatter. Barker & Goldstein [3] and Barker and Pawlewicz [4] have developed empirical equations relating %Ro to temperature (T °C):

$$T \,°C = [(\ln \%Ro) + 1.26] / 0.0081 \quad (1)$$
$$T \,°C = [(\ln \%Ro) + 1.68] / 0.0124 \quad (2)$$
$$T \,°C = [(\ln \%Ro) + 1.19] / 0.00782 \quad (3)$$

The authors [4] suggested that equation (1) is of general applicability, but that equation (2) is more appropriate for burial heating and equation (3) for hydrothermal metamorphism.

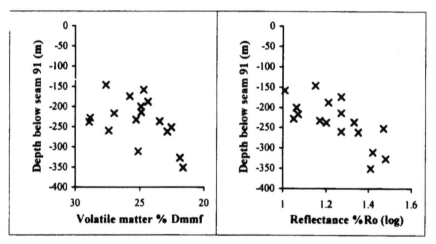

Figure 4. Composite plots of rank parameters against depth below the No2 Rhondda seam, (seam 91): a) %Vm$_{dmmf}$, b) %Ro. Collieries used are locations 2, 3, 5, 6, 8, 9, 11 and 12 in figure 1.

Using equation (1), the increase in temperature with depth has been calculated (Fig. 4c). The linear regression of these data suggests a palaeogeothermal gradient of approximately 218 °C km^{-1}. Using the burial heating equation (2) the palaeogeothermal gradient is approximately 182 °C km^{-1}. However, the scatter in the data makes the resulting gradients very uncertain, with a range from a minimum value of approximately 115 °C km^{-1} to a maximum of approximately 320 °C km^{-1}. This scatter may be partly the result of the geographical spread of the sampling points, which was caused by the availability of seams for which both %Vm$_{dmmf}$ and %Ro data had been determined. However, in all these sampling areas the No2 Rhondda seam is thought to have had the same rank.

Excursions from Hilt's Law

The scatter in maturity parameters plotted against depth (Figs. 4a, b & c) can be investigated by analysing the data from a sequence of coal seams within one colliery. Using data from a group of three adjacent collieries in the north-east of the coalfield (locations in Fig. 1) the variation in raw volatile matter, %Vm$_{raw}$ and %Vm$_{daf}$ with depth for 10 coal seams was plotted (Fig. 5). In all three collieries a reversal in coal rank increase with depth (an 'excursion') occurs between the Lower Six Feet (seam 7) and the Upper Nine Feet (seam 6) independently of the absolute value of %Vm$_{daf}$. Thus the excursion is from 15 and 17 %Vm$_{daf}$ at Groesfaen Colliery, between 16 and 18 %Vm$_{daf}$ at Bargoed Colliery and between 20 and 25 %Vm$_{daf}$ at Oakdale Colliery, reflecting the regional decrease in rank from west to east in this part of the coalfield.

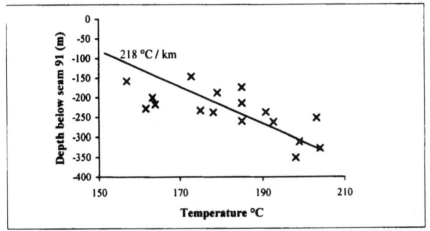

Figure 4c. Calculated temperature against depth below the No2 Rhondda seam, (seam 91). Collieries used are locations 2, 3, 5, 6, 8, 9, 11 and 12 in figure 1.

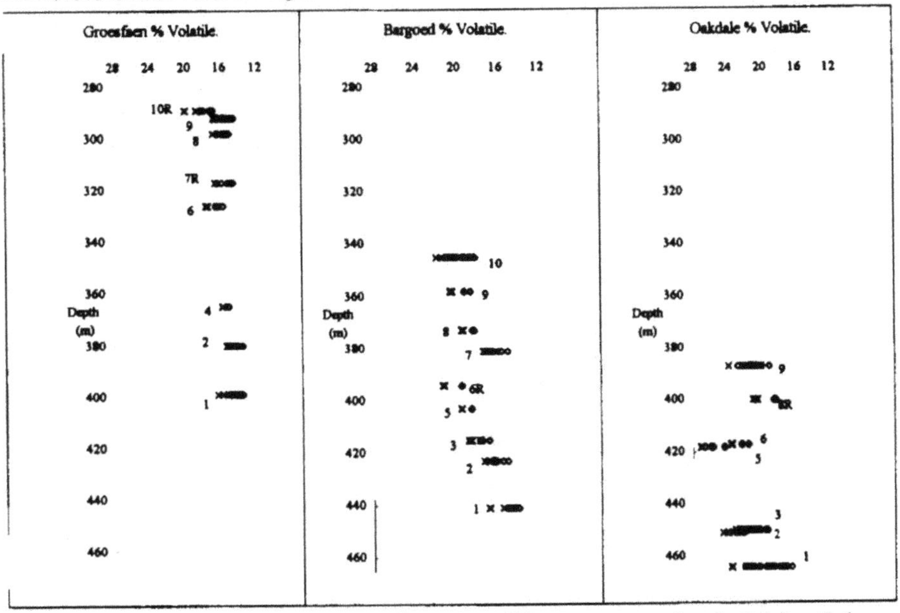

Figure 5. Groesfaen to Oakdale section, Vm$_{raw}$, diamonds and Vm$_{daf}$, crosses, against depth below Ordnance Datum. Seams are 1 Gellideg; 2 Five Feet; 3 Seven Feet / Yard; 4 Amman Rider; 5 Lower Nine Feet; 6 Upper Nine Feet; 7 Lower Six Feet; 8 Upper Six feet; 9 Four Feet; 10 Two Feet Nine. R denotes rashings horizons.

R. Gayer and R. Fowler

Similar excursions in %Vm_daf from Hilt's Law were recorded by Gayer et al. [17] but at different stratigraphic levels in different collieries. The combined effect of these excursions would be to produce the scatter shown in the data sets (Fig. 4a).

Insufficient vitrinite reflectance measurements are available in the British Coal archives to carry out a similar analysis of %Ro with depth in an individual colliery. As all but one of the deep mines in the coalfield have closed, it is no longer possible to sample directly from the collieries. However, we have obtained coal samples from Wyllie Colliery (located in Fig. 1) from the Wyllie collection (78.50.G) in the National Museum of Wales in Cardiff. %Ro of telocollinites from these samples were measured using procedures which meet the relevant parts of ISO 7404. The results (Table 1 and Fig. 6) show three excursions from Hilt's Law; the first between the No2 Rhondda and the Pentre Rider seams, the second between the Upper Two Feet Nine and the Upper Six Feet seams, and the third between the Bute and the Amman Rider seams. A range of gradients in %Ro is indicated, mirroring the scatter in values seen in the composite plot (Fig. 4b). The equivalent temperature gradients, using the equation 1 [3], are 95 °C km^{-1} and 309 °C km^{-1}, which are comparable to the range noted above (Fig. 4c).

Table 1. Wyllie Colliery Petrological data of sampled seams.

Seam Name	Sample	% Ro	Standard Dev	Vitrinite	Liptinite	Inertinite	Pyrite	Other
Upper No2 Rhondda	91	0.92	0.020	70.94%	5.77%	18.59%	0.43%	4.27%
Lower No2 Rhondda	95	1.16	0.020	78.30%	4.47%	14.47%	0.43%	2.34%
Pentre Rider	102	1.02	0.029	41.28%	9.50%	12.02%	0.58%	36.63%
Pentre	103	1.12	0.020	70.42%	7.04%	17.37%	0.70%	4.46%
Upper Two feet nine	106	1.23	0.025	38.68%	7.08%	42.45%	2.83%	8.96%
Lower 2'9" or Four Feet	109	1.09	0.020	68.28%	6.30%	15.97%	0.42%	9.03%
Lower 2'9" or Four Feet	107	0.96	0.020	69.08%	7.25%	17.87%	0.48%	5.31%
Upper Six Feet	114	0.90	0.020	69.89%	5.68%	20.42%	0.21%	3.79%
Middle Six Feet deformed	115	1.17	0.024	40.21%	6.25%	39.17%	1.04%	13.33%
Lower Six Feet	116	1.04	0.025	37.73%	6.82%	45.45%	0.45%	9.55%
Upper Nine Feet	121	1.29	0.021	52.24%	5.88%	34.59%	0.24%	7.06%
Bute	129A	1.37	0.020	64.97%	6.26%	23.48%	0.20%	5.09%
Amman Rider	131B	1.06	0.020	71.81%	5.78%	20.72%	0.24%	1.45%
Yard	131A	1.47	0.022	62.02%	18.75%	16.59%	0.24%	2.40%
Seven Feet	131C	1.27	0.025	58.31%	22.25%	17.10%	0.23%	2.11%

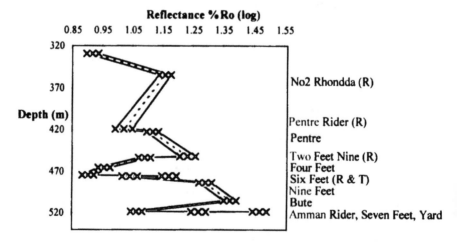

Figure 6. Wyllie Colliery %Ro against depth, dashed line is the mean, solid lines are ± 1 standard deviation. T denotes thrust and R denotes rashings horizons.

Variation of coal rank parameters within seam profiles

In order to investigate the variation in %Ro within a seam profile, a coal sequence exposed in Llanilid West Opencast Coal Site (located in Fig. 1) was logged and individual coal seams sampled by channelling. Each channel was divided into mainly 10 cm lengths and each length separately bagged. The samples were prepared and petrographically analysed using the relevant parts of ISO 7404 (Table 2). The values of %Vm_{daf} (analyses provided by Celtic Energy Ltd.) and %Ro were plotted against depth respectively (Figs.7a & b) and both show coincident excursions from Hilt's Law between the Six Feet and Nine Feet seams, the Bute and Amman Rider seams and between the Yard and Gellideg seams, producing a scatter of data similar to those described above. However, the depth scale in figure 7 makes it difficult to see the fine scale variations of %Ro within each seam. The same data are shown at a larger scale and without the intervening non-coal sequence in Fig. 8a. Major and statistically significant variations in %Ro are apparent within the profile of each seam. Comparison of the variations in %Ro with the maceral group composition of the same samples (Fig. 9) and with the Tissue Preservation Index (TPI) and Gellification Index (GI) (Figs. 10a & b) show no apparent statistical correlation.

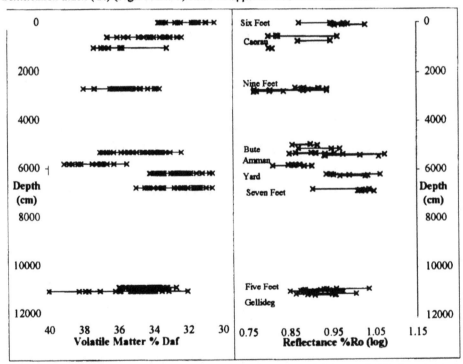

Figure 7. Llanilid West Opencast variation in rank parameters against depth below roof of Six Feet seam (seam 37): a) %Vm_{daf}, b) %Ro, mean only.

Correlation between variation in coal rank parameters and deformation

In order to investigate any possible correlation between coal seam deformation and %Ro, a quantitative measure of the deformation is required. Existing measures for small coal samples, such as the Average Structural Index (ASI) [11] and Modified Structural Index (MASI) [5], proved difficult to apply in the field to coals with low levels of deformation. A new index, called the Numerical Deformation Index (NDI), has been designed to quantify numerically small-scale deformation which is applicable to small coal samples (up to 10 cm^3). The NDI sums values assigned to each of three aspects of coal deformation: i) preservation of cleat fractures, ii) degree

R. Gayer and R. Fowler

Table 2.   Llanilid West Opencast Petrological and NDI data of sampled seams.

Seam Name	Sample	% Ro	Standard Dev	Vitrinite	Liptinite	Inertinite	Pyrite	Other	NDI
Six Feet	133J	0.87	0.023	33.59%	37.38%	15.75%	0.57%	12.71%	13
	133I	0.98	0.055	59.27%	15.68%	18.93%	0.19%	5.93%	14
	133H	0.95	0.042	50.20%	14.55%	28.48%	0.41%	6.35%	14
	133G	0.95	0.022	37.70%	36.07%	18.03%	0.55%	7.65%	16
	133F	0.99	0.023	65.86%	15.21%	14.29%	0.00%	4.64%	14
	133E	0.95	0.021	41.77%	34.01%	9.80%	0.92%	13.49%	15
	133D	0.98	0.023	52.91%	28.29%	7.36%	1.16%	10.27%	15.5
	133C	0.97	0.061	39.35%	38.71%	12.90%	1.13%	7.90%	13
	133B	1.03	0.044	29.41%	50.42%	16.81%	0.34%	3.03%	18
	133A	0.96	0.023	45.04%	32.25%	7.63%	0.76%	14.31%	16
Upper Caerau	122E	0.82	0.025	51.04%	33.90%	12.05%	0.75%	2.26%	6
	122D	0.82	0.024	50.28%	32.97%	11.79%	0.55%	4.42%	6
	122C	0.80	0.023	50.74%	30.15%	13.60%	0.74%	4.78%	6
	122B	0.96	0.033	58.35%	25.72%	11.71%	0.19%	4.03%	8
Lower Caerau	119B	0.95	0.035	59.77%	20.30%	9.77%	4.32%	5.83%	9
	119A	0.87	0.026	57.79%	20.83%	11.63%	2.06%	7.69%	8
Red	110B	0.80	0.028	48.36%	29.82%	15.09%	0.73%	6.00%	6
	110A	0.81	0.040	54.50%	19.27%	15.23%	1.10%	9.91%	6
Nine Feet	30L	0.86	0.029	61.36%	24.47%	11.07%	0.19%	2.91%	7
	30K	0.88	0.027	71.46%	16.89%	7.96%	0.58%	3.11%	6.5
	30J	0.92	0.027	63.85%	25.38%	8.46%	0.58%	1.73%	7
	30I	0.92	0.028	63.58%	24.66%	10.02%	0.19%	1.54%	6
	30H	0.88	0.029	50.68%	34.82%	12.96%	0.19%	1.35%	6
	30G	0.94	0.028	59.67%	26.39%	12.45%	0.19%	1.30%	7
	30F	0.88	0.023	45.62%	42.27%	9.12%	0.37%	2.61%	7
	30E	0.89	0.023	54.04%	33.85%	8.85%	0.38%	2.88%	8
	30D	0.77	0.029	59.67%	15.24%	21.56%	0.74%	2.79%	6
	30C	0.80	0.029	68.86%	9.09%	8.32%	3.68%	10.06%	7
	30B	0.94	0.028	60.42%	26.89%	10.80%	0.76%	1.14%	9
	30A	0.88	0.030	52.26%	30.57%	15.09%	0.38%	1.70%	6
	30N	0.84	0.046	48.60%	15.83%	27.00%	3.35%	5.21%	9
	30O	0.80	0.046	49.53%	16.13%	26.19%	4.17%	3.98%	10
	30P	0.77	0.046	50.19%	17.62%	26.44%	1.72%	4.02%	10
	30Q	0.77	0.045	40.11%	13.10%	22.46%	2.32%	21.93%	10
Upper Bute Rd.	44A	0.90	0.034	56.47%	24.58%	12.76%	4.32%	1.88%	9.5
	44B	0.90	0.029	59.61%	24.14%	11.58%	2.96%	1.72%	10
	44C	0.86	0.028	58.52%	25.28%	12.78%	1.42%	1.99%	10
	44D	0.92	0.029	64.74%	17.88%	11.34%	4.79%	1.26%	10
Lower Bute Rd.	46A	0.87	0.028	57.43%	26.73%	12.13%	1.98%	1.73%	10
	46B	0.95	0.029	66.21%	15.52%	11.59%	3.34%	3.34%	11
	46C	0.97	0.028	59.79%	25.26%	11.34%	1.29%	2.32%	11.5
	8	0.88	0.029	94.00%	5.64%	0.36%	0.00%	0.00%	7
Bute	48A	0.90	0.038	59.87%	14.93%	9.78%	3.98%	11.44%	10
	48B	0.91	0.041	63.48%	20.31%	9.04%	1.02%	6.14%	10
	48C	0.86	0.044	46.62%	16.05%	8.95%	2.20%	26.18%	8.5
	48D	0.91	0.042	64.01%	16.30%	11.38%	1.87%	6.45%	9
	48E	0.85	0.042	63.77%	15.46%	11.11%	2.90%	6.76%	9
	48F	0.95	0.041	69.58%	15.05%	10.36%	2.43%	2.59%	11
	48G	0.93	0.042	68.20%	18.20%	8.36%	0.49%	4.75%	10
	48H	0.97	0.042	65.95%	18.18%	8.26%	2.48%	5.12%	10

Seam Name	Sample	% Ro	Standard Dev	Vitrinite	Liptinite	Inertinite	Pyrite	Other	NDI
	48I	1.07	0.044	47.46%	13.90%	7.46%	0.85%	30.34%	13.5
	48J	1.02	0.041	62.91%	20.68%	9.91%	1.03%	5.47%	13
	48K	0.93	0.043	58.23%	13.58%	22.41%	3.06%	2.72%	9
	48L	0.93	0.042	65.88%	13.34%	15.37%	2.53%	2.87%	9
	48N	1.06	0.042	68.20%	16.72%	10.33%	1.80%	2.95%	12
Amman Rider	60A	0.87	0.022	58.70%	18.70%	9.07%	10.74%	2.78%	10.5
	60B	0.86	0.024	55.93%	21.11%	8.52%	13.15%	1.30%	10
	60C	0.85	0.022	60.64%	13.83%	8.87%	14.54%	2.13%	10
	60D	0.86	0.024	70.02%	10.40%	9.01%	9.01%	1.56%	10
	60E	0.81	0.022	47.96%	35.92%	8.93%	5.83%	1.36%	9
	60F	0.90	0.024	50.29%	34.68%	8.86%	4.82%	1.35%	11
	60G	0.88	0.027	72.65%	10.78%	11.78%	2.40%	2.40%	11
Yard	70A	0.93	0.023	44.74%	41.53%	11.23%	0.89%	1.60%	13
	70B	0.95	0.036	37.75%	30.02%	17.86%	0.55%	13.81%	12
	70C	1.06	0.048	41.88%	25.65%	15.13%	0.92%	16.42%	16
	70D	0.95	0.042	55.81%	18.22%	18.22%	0.58%	7.17%	12
	70E	0.99	0.037	36.75%	38.39%	17.73%	0.55%	6.58%	13.5
	70F	1.03	0.032	38.10%	36.25%	14.68%	0.56%	10.41%	16
	70G	0.97	0.044	64.03%	18.07%	14.80%	0.69%	2.41%	12
	71,	1.03	0.060	58.03%	17.01%	16.64%	0.19%	8.13%	15
Seven Feet	82K	0.90	0.029	61.47%	20.79%	10.57%	3.05%	4.12%	12
	82L	1.04	0.030	70.15%	14.33%	11.47%	0.51%	3.54%	15
	82M	1.03	0.037	58.56%	18.82%	16.54%	0.95%	5.13%	14
	82N	1.01	0.028	61.58%	18.77%	14.74%	0.53%	4.39%	15
	82O	1.01	0.028	45.40%	15.95%	26.64%	0.56%	11.44%	15
	82P	1.01	0.030	50.62%	24.08%	21.09%	0.53%	3.69%	15
	82Q	1.03	0.035	55.44%	14.51%	25.91%	0.35%	3.80%	15
	82R	1.05	0.038	60.60%	19.44%	13.31%	1.23%	5.43%	16
Five Feet	88O	0.96	0.051	47.67%	29.21%	20.07%	0.54%	2.51%	12
	88N	0.93	0.052	49.01%	27.11%	18.31%	3.59%	1.97%	10.5
	88M	1.03	0.044	72.08%	14.38%	8.80%	3.38%	1.35%	14
	88L	0.93	0.044	72.33%	11.38%	10.36%	3.40%	2.55%	12
	88K	0.88	0.045	66.23%	21.36%	6.62%	3.31%	2.48%	12
	88J	0.92	0.031	75.61%	8.44%	13.51%	0.75%	1.69%	10
	88I	0.88	0.048	57.85%	14.81%	24.51%	0.88%	1.94%	9
	88H	0.94	0.049	50.46%	25.32%	17.85%	2.55%	3.83%	9.5
	88G	0.87	0.049	49.91%	22.75%	21.34%	1.94%	4.06%	8
	88F	0.91	0.044	73.39%	9.15%	12.54%	1.02%	3.90%	8.5
	88E	0.85	0.036	47.90%	12.79%	31.68%	0.57%	7.06%	9
	88D	0.90	0.041	72.24%	14.21%	11.54%	0.67%	1.34%	8
	88C	0.97	0.044	72.57%	11.41%	12.27%	0.85%	2.90%	9
	88B	0.97	0.044	70.18%	14.00%	9.88%	0.99%	4.94%	12
	88A	0.95	0.044	71.23%	19.69%	6.85%	0.51%	1.71%	12
Gellideg	86L	0.88	0.044	67.95%	16.46%	10.86%	2.80%	1.93%	10.5
	86K	0.91	0.052	52.35%	26.08%	14.31%	5.49%	1.76%	12
	86J	0.90	0.027	59.36%	28.65%	8.99%	2.25%	0.75%	10
	86I	0.89	0.024	57.72%	31.80%	6.62%	3.13%	0.74%	12
	86H	0.91	0.044	71.11%	19.66%	6.84%	1.03%	1.37%	12
	86G	1.00	0.030	66.35%	13.62%	18.27%	1.12%	0.64%	14
	86F	0.86	0.044	72.79%	10.85%	12.68%	1.65%	2.02%	12
	86E	0.95	0.048	59.33%	13.91%	24.47%	0.35%	1.94%	12.5
	86D	0.89	0.030	39.76%	14.51%	44.71%	0.34%	0.68%	11
	86C	0.96	0.052	48.19%	25.99%	14.80%	3.07%	7.94%	16
	86B	0.95	0.030	36.71%	16.78%	35.88%	1.99%	8.64%	14.5
	86A	0.91	0.030	69.44%	13.92%	14.65%	0.36%	1.63%	13

of competence/coherence and iii) presence of deformation structures. Both i) and ii) are assigned values from 1 - 5, whilst iii) has values from 3 -15, according to the descriptions in Table 3. Thus the NDI ranges from 5 for a non-deformed coal to 25 for the highest degree of deformation. However, in very highly sheared coals, most of the criteria used to assign an NDI may have been destroyed. In these coals it is important to combine data from both large- and small-scale observations to arrive at an accurate value of NDI.

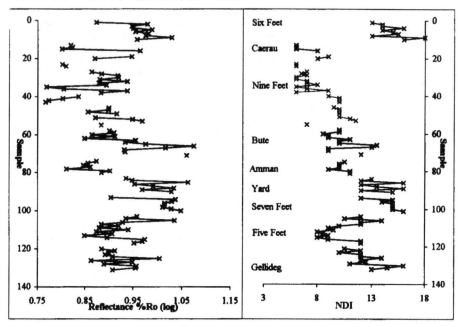

Figure 8. Llanilid West Opencast plots showing in seam variations of a) %Ro and b) NDI. Lines join mean values within the same seam.

Table 3. Outline descriptions of the criteria used to compile the Numerical Deformation Index.

i) Cleat
perfect cubic cleat, all perpendicular	1
some cleat, or slightly inclined	3
no cleat, completely destroyed	5

ii) Competence and coherence, resistance to disintegration under normal handling conditions after drying
large block, 25 cm x 15 cm x 10 cm	1
blocky, 10 cm	2
powdery / rubbly, mixture of pebbles and powder	3
friable, beans or nut size, 1 cm	4
complete lack of cohesiveness, duff	5

iii) Deformation

folds and faults		fabrics	
plies straight and undeformed	1	no physically defining deformation fabric	1
minor faulting	3	rhombic blocks of coal caused by slip	2
folded and deformed	5	surfaces which show signs of movement or listric	3
		locally intense deformation structures, multi slip planes	4
		flower structures	5

angle of inclination of slip or shear surfaces
0 - 45°	5
45 - 75°	3
75 - 90°	1

The values of NDI were determined for the same 10 cm sample lengths of coals at Llanilid West Opencast coal site as those used for petrological investigation (Fig. 8b). To test the apparent correlation between %Ro and NDI (Fig. 8), the two sets of data for each coal seam were normalised to percentages and statistical correlation levels determined for each coal seam. The results (Fig. 11) show that the R^2 correlation index ranges from 0.185 to 0.995, with good correlations in the Seven Feet (0.82), Yard (0.85), Amman Rider (0.945) and Bute (0.78) seams.

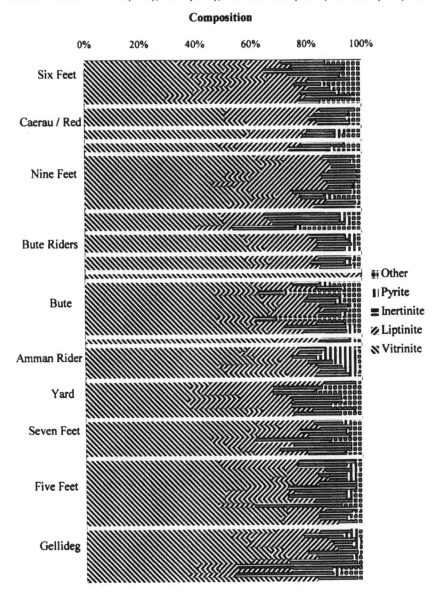

Figure 9. Llanilid West Opencast Maceral group composition against seam sample.

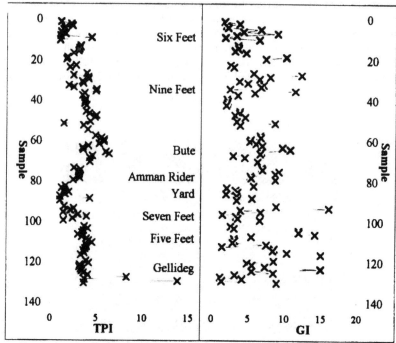

Figure 10. Llanilid West Opencast plots showing in seam variations in petrographic indices a) TPI, b) GI.

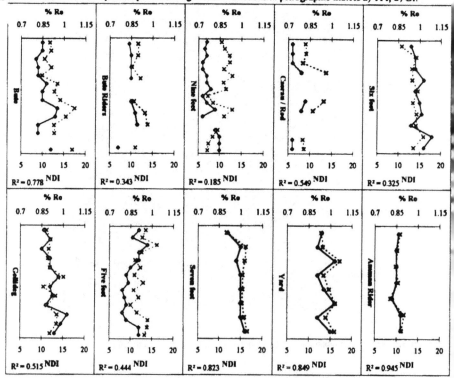

Figure 11. Llanilid West Opencast seam correlations between %Ro, solid/diamond and NDI, dash/cross.

MATURITY PROFILE MODELLING

The coal rank data presented in this paper suggest that the average palaeogeothermal gradient through the Westphalian Coal Measures in the east of the coalfield was approximately 200 °C km⁻¹ at the time of rank development during the Variscan deformation. The scatter of data about this gradient could indicate a possible range in palaeogeothermal gradient from approximately 95 °C km⁻¹ to approximately 320 °C km⁻¹. The data from individual collieries demonstrate the presence of excursions from Hilt's Law which would account for the scatter in composite rank versus depth plots.

We have attempted to model this pattern of maturity variation with depth to establish the likely sources of heat in the foreland basin. The modelling used a 1D basin modelling PC package, BasinMod® version 5.40 which requires that each unit in a stratigraphic sequence, representing the basin-fill and its basement, is assigned a series of parameters. These parameters control the passage of fluids and heat through the section as the sequence is deposited and compacted, and allow the temperatures and maturity gradients to be calculated for any time slice during the basin evolution. Table 4 shows the stratigraphy and Table 5 shows the main lithological parameters used in the modelling.

The sources of heat in the modelling can be heat flow rising from the underlying basement (basal heat flow), heat generated within the section from radiogenic sources, and heat flowing laterally into the section via fluids or igneous intrusions. By varying these heat sources within the limits dictated by hypotheses of basin evolution, the modelling attempts to match the maturity gradients calculated in the model to those measured in the rocks.

We have used a value of 63 mWm⁻² for the basal heat flow, based on the heat flow in a peripheral foreland basin on established, normal thickness continental crust [1]. Using only this heat source resulted in a model (Fig. 12), in which the calculated maturity levels in the Coal Measures range from 0.4 - 0.7 %Ro, too low to match the observed maturity levels of 0.7 - 1.05 %Ro. By allowing hot fluids to flow into the permeable underlying Carboniferous limestones and the interbedded Six Feet sandstone a good model match was obtained for the maturity levels in the sampled coal seam sequence (Fig. 13). However, the maturity at the top of the Westphalian D is too low.

We thus conclude that the coal rank was developed by the following heat sources:
a) a basal heat flow of 63 mWm⁻²;
b) a lateral heat flow of :
    i) approximately 35 mWm⁻² into the underlying Carboniferous limestones;
    ii) approximately 1 mWm⁻² into the Six Feet sandstone;
    iii) an as yet undetermined amount into the Upper Westphalian sequence.

This model does not require the former existence of a thick cover sequence, subsequently eroded and for which there is little evidence [37], nor does it require the presence of abnormally high basal heat flow levels, which are inappropriate for a peripheral foreland basin setting [27]. The model is in accordance with the preliminary results from 2-D and 3-D basin-scale modelling of fluid flow through the foreland basin, which uses finite element techniques and is currently being carried out by R. Gayer & G. Garven.

Table 4. Stratigraphy used for BasinMod© modelling. Symbols for Type are F, Formation; E, Erosion and D, Deposition.

Period	Formation / Event Name	Type & Begin Age	Well Top	Present Thickness	Missing Thickness	Lithology
Triassic	Triassic Penarth Group	F 240	0	10		Sandstone
	Mercia Mudstone	F 245	10	70		Shale
Carboniferous	Stephanian removal	E 290			-1430	Sandy muds
	Unconformity 1	E 295			-400	Sandstone
Stephanian	Stephanian	D 303			400	Sandstone
Westphalian D	Westphalian D	D 307			1170	Sandy muds
	Llanharan thrust repeat	D 307.5			260	Coal measures
Westphalian C	Uppermost Lynfi West C	F 308	80	8		Siltstone
	No 3 Rhondda	F 308.7	88	1		Coal
	Lynfi beds Sandstone	F 309	89	42		Sandstone
	Hafod	F 309.4	131	1		Coal
	Mid Westphalian C	F 309.8	132	39		Shale
	Abergorki	F 310	171	0.4		Coal
	Mid West C muds	F 310.1	171.4	7		Shale
	Mid West C sands	F 310.2	178.4	15		Sandstone
	Mid West C lower muds	F 310.3	193.4	6		Shale
Westphalian B	Pentre group	F 310.4	199.4	2		Coal
	Interseam muds	F 310.5	201.4	3.5		Shale
	Gorllwyn	F 310.6	204.9	2		Coal
	Gorllywn Sandstone	F 310.7	206.9	2		Sandstone
	Lowest C muds and sands	F 310.8	208.9	30		Silty mudstone
	Uppermost B silts	F 311	238.9	22		Siltstone
	Two feet nine	F 311.1	260.9	2		Coal
	Mudstone seatearth	F 311.2	262.9	3		Shale
	Four feet	F 311.3	265.9	2		Coal
	12m mudstone	F 311.4	267.9	12		Shale
	Six feet Sandstone	F 311.5	279.9	10		Sandstone
	Six feet	F 311.7	289.9	1		Coal
	Caerau mudstones	F 311.8	290.9	8		Shale
	Caerau / Red group	F 311.9	298.9	1		Coal
	18m mudstone	F 312	299.9	18		Shale
	Nine feet	F 312.1	317.9	2		Coal
	Lower nine feet muds	F 312.3	319.9	21		Shale
	3m mudstone	F 312.4	340.9	3		Shale
	Bute	F 312.5	343.9	1.6		Coal
	3.2m mudstone and sand	F 312.6	345.5	3.2		Silty mudstone
Westphalian A	Amman rider	F 312.7	348.7	1		Coal
	intermudstone	F 312.8	349.7	3.3		Sandy muds
	Yard	F 312.9	353	1		Coal
	5m yard mudstone	F 313	354	5		Shale
	Seven feet	F 313.3	359	1		Coal
	40m mudstone	F 313.4	360	40		Shale
	Gellideg	F 313.6	400	2		Coal
	42m mudstone	F 313.8	402	42		Shale
	Garw	F 314	444	1		Coal
	Garw mudstone and sands	F 314.2	445	95		Sandy muds
	Cefn Cribbur Sandstone	F 314.4	540	30		Sandstone
	140m muds and sands	F 314.6	570	140		Sandy muds
Namurian	Millstone Grit	F 315	710	200		Millstone grit
Dinantian	Oxwich head limestone	F 322	910	130		Limestone
	Pant mawr Sandstone	F 323	1040	3		Sandstone
	Hunts bay oolite	F 325	1043	150.25		Dolomite
	High Tor	F 330	1193.25	60		Limestone
	Caswell Bay mudstone	F 335	1253.25	10		Shale
	Gully oolite	F 340	1263.25	40		Dolomite
	Friars point limestone	F 345	1303.25	80		Limestone
	B'y Bofiscin limestone	F 350	1383.25	50		Limestone
	Cwmniscay mudstone	F 352	1433.25	50		Shale
	Castell Coch limestone	F 355	1483.25	25		Limestone
	Tongwynlais formation	F 360	1508.25	45		Dolomite
Devonian	Quartz Conglomerate	F 365	1553.25	60		Sandstone
	Cwrt yr ala Formation	F 370	1613.25	30		Sandstone
	Lower Devonian brownst	F 390	1643.25	70		Sandstone
	Llanishen Conglomerate	F 400	1713.25	10		Sandstone

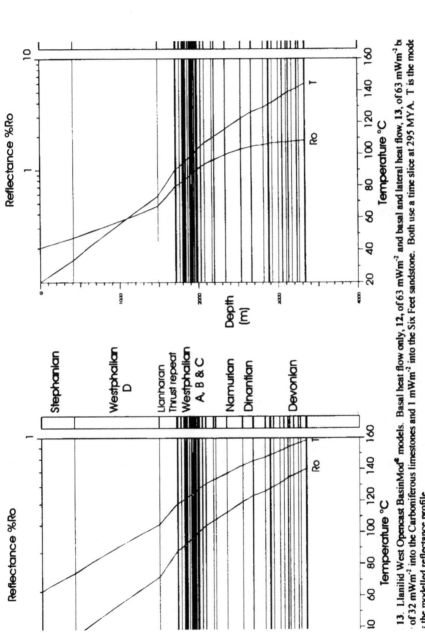

13. Llanilid West Opencast BasinMod® models. Basal heat flow only, 12, of 63 mWm⁻² and basal and lateral heat flow, 13, of 63 mWm⁻² b of 32 mWm⁻² into the Carboniferous limestones and 1 mWm⁻² into the Six Feet sandstone. Both use a time slice at 295 MYA. T is the mode s the modelled reflectance profile.

Table 5. Lithological parameters used for BasinMod® modelling.

Lithology Name	Composition Sandstone	Silt	Shale	Limestone	Dolomite	Evaporite
Sandstone	100	0	0	0	0	
Millstone grit	30	30	40	0	0	
Sandy muds	48	0	47	0	0	
Siltstone	0	100	0	0	0	
Silty mudstone	0	33	67	0	0	
Shale	0	0	100	0	0	
Limestone	0	0	0	100	0	
Dolomite	0	0	0	0	100	
Evaporite	0	0	0	0	0	100
Coal	0	0	0	0	0	0
Igneous	0	0	0	0	0	0
Coal measures	30	20	25	0	0	0

Lithology Name	Composition Kerogen	Igneous	Heat Capacity kJ/m³/°C	Conductivity Correction	Conductivity W/m/°C	Density g/cm³
Sandstone	0	0	2800	270	4.4	2.64
Millstone grit	0	0	2475	60	2.258	2.624
Sandy muds	5	0	2378	57.5	2.569	2.579
Siltstone	0	0	2650	170	2	2.64
Silty mudstone	0	0	2281	-64.5	1.649	2.613
Shale	0	0	2100	-180	1.5	2.6
Limestone	0	0	2600	350	2.9	2.72
Dolomite	0	0	2600	300	4.8	2.85
Evaporite	0	0	1750	470	5.4	2.15
Coal	100	0	950	250	0.3	1.8
Igneous	0	100	2500	380	2.9	2.65
Coal measures	25	0	1832	173.5	0.955	2.256

DISCUSSION

The maturity modelling, described above, has been able to explain the overall increase in rank with depth recorded in the south-east of the coalfield. However, the excursions from Hilt's Law demonstrated in the %Vm$_{daf}$ data (Figs. 5, 7a) and in the %Ro data (Figs. 6, 7b) have, as yet not been modelled. It is probable that these excursions, which involve several coal seams in the sequence, are the result of fluid flow carrying heat laterally into the coals. Unlike the lateral heat flow through the Carboniferous limestones and the Six Feet sandstone, the lateral heat flow into the coal seams did not produce an equilibrated temperature profile. It is likely that this is due to the short duration of fluid pulses which would have increased the temperature of the coal seam through which the fluid migrated and of the immediately adjacent sequence, but would not have allowed a complete relaxation of the thermal profile. One possible explanation for such a pulsed fluid flow would be seismic pumping [31] associated with movements along the thrust detachments that have been observed in the relevant coals (Figs. 5, 6).

Although the major variations in coal rank with depth described in this paper can be explained by a combination of basal heat flow and lateral, fluid-hosted heat flow, the detailed variations in rank through a single seam profile cannot be attributed to heat carried into the seam by fluids. It is possible that such fluids could be guided through the different plies of a coal by virtue of their increased fracture permeability resulting from the deformation fabrics, but the consequent ncrease in temperature would rapidly equilibrate through the seam. Using the values of heat conductivity and heat capacity for coal, a seam 1 m thick with a 10 cm thick hot fluid conduit at ts base should reach a thermal equilibrium within a few seconds. Thus no variation in rank would be expected.

Several cases of suppression of vitrinite reflectance in coals have been reported [29, 34]. Mukhopadhyay [28] reviewed the possible causes for vitrinite reflectance suppression most of which were associated with the original depositional environment of the coal-forming peat or with retention of bitumens or migrated oil. Our observations do not appear to indicate any relationship to depositional setting; there is no correlation between %Ro and TPI/GI. However, there is a need to investigate the possibility of a link between %Ro and hydrocarbon generation and retention in these South Wales coals.

The strong correlation between variation in vitrinite reflectance and deformation suggests that vitrinite reflectance may be affected by shear strains associated with thrust deformation. Various studies have attempted to establish such a link [8, 39], but no unambiguous relationships have resulted. It is possible that shear strains would increase the vitrinite reflectance anisotropy, and the study of Salih & Lisle [30] has shown such a relationship in folded anthracites in the South Wales coalfield, although the results related to the magnitude of the anisotropy and not to the absolute value of vitrinite reflectance. Further work on the effects of shear strain on vitrinite reflectance is planned.

CONCLUSIONS

Coals in the South Wales coalfield show an increase in rank not only with stratigraphical depth, obeying Hilt's law, but also laterally towards the north-west of the coalfield. The rank was developed both before and during Variscan thrust deformation.

Plots of maturity parameters against depth show a wide scatter, but indicate a palaeogeothermal gradient of around 218 °C km^{-1} within the Westphalian Coal Measures in the eastern part of the coalfield.

Excursions from Hilt's law occur in locally developed zones associated with one or more coal seams, and account for the scatter of maturity parameters. These excursions are interpreted as localised zones of higher temperature, associated with increased thrust deformation and are possibly generated by seismic pumping of hot fluids along active thrusts. The palaeogeothermal gradients associated with the excursions are approximately 320 °C km^{-1}.

Preliminary basin modelling using BasinMod® software suggests a basal heat flow of 63 mWm^{-2} and lateral heat flow into the underlying Carboniferous limestones of approximately 32 mWm^{-2} and into the Six Feet sandstone of approximately 1 mWm^{-2}.

Detailed variations in %Ro in some seam profiles are correlated with a numerical deformation index (NDI) and may indicate that shear strain contributes to the development of vitrinite reflectance.

Acknowledgements.

We are grateful to British Coal Opencast and to Celtic Energy for allowing access to opencast coal mines in South Wales, for providing proximate analysis data, and the plans and sections from which the thrust structure within the South Wales coalfield has been deduced. We are also grateful to the National Museum of Wales who kindly provided the coal samples from Wylli Colliery and provided technical assistance. We thank Platte River Associates for loaning us the BasinMod© computer package.

REFERENCES

1. Allen, P.A. & Allen, J.R. *Basin Analysis: Principles and Applications.* Blackwell Scientific Publications, Oxford (1990).
2. Austin, R.L. & Burnett, R.D. Preliminary Carboniferous conodont CAI data South Wales, the Mendips and adjacent areas, United Kingdom. *Mémoires Institut Géologique de l'Université Catholique de Louvain*, **35**, 137-153 (1994).
3. Barker, C.E. & Goldstein, R.H. Fluid inclusion technique for determining maximum temperature in calcite and its comparison to the vitrinite reflectance geothermometer. *Geology*, **18**, 1003-1006 (1990).
4. Barker, C.E. & Pawlewicz, M.J. Calculation of vitrinite Reflectance from Thermal Histories and Peak Temperatures. A Comparison of Methods. In: *Vitrinite Reflectance as a Maturity Parameter, Applications and Limitations*, P.K. Mukhopadhyay and W.G. Dow (Eds). pp. 218 229. American Chemical Society, ACS Symposium series **570**, Washington DC (1994).
5. Barker-Read, G.R. *The geology and related aspects of coal and gas outbursts in the Gwendraeth Valley* [unpublished MSc thesis] University of Wales, Cardiff (1980).
6. Bevins, R.E., White, S.C. & Robinson, D. The South Wales Coalfield: low grade metamorphism in a foreland basin setting? *Geological Magazine*, **In Press** (1996).
7. Brooks, M., Miliorizos, M. & Hillier, B.V. Deep structure of the Vale of Glamorgan, South Wales, UK. *Journal of the Geological Society, London*, **151**, 909-917 (1994).
8. Bustin, R.M. Heating during thrust faulting in the Rocky Mountains: Friction or Fiction? *Tectonophysics*, **95**, 309-328 (1983).
9. Cole, J.E., Miliorizos, M., Frodsham, K., Gayer, R.A., Gillespie, P.A., Hartley, A.J. & White, S.C. Variscan structures in the opencast coal sites of the South Wales Coalfield. *Proceedings of the Ussher Society*, **7**, 375-379 (1991).
10. Fermont, W.J.J. Possible causes of abnormal vitrinite reflectance values in paralic deposits of the Carboniferous in the Achterhoek area, The Netherlands. *Organic Geochemistry*, **12/4**, 401-411 (1988).
11. Fisher, R.J. *A study of variations in the physical characteristics of West Wales anthracite* [unpublished PhD thesis] University of Wales, Cardiff (1977).
12. Frodsham, K., Gayer, R.A., James, J.E. & Pryce, R. Variscan thrust deformation in the South Wales Coalfield - a case study from Ffos-Las Opencast Coal Site. In: *The Rhenohercynian and Subvariscan Fold Belts*. R.A. Gayer, R.O. Greiling, & A. Vogel (Eds). pp. 316-348. Earth Evolution Science Series, Vieweg, Braunschweig (1993).
13. Frodsham, K. & Gayer, R.A. Variscan compressional structures within the main productive coal bearing strata of South Wales. *Journal of the Geological Society, London*, **154, In Press** (1997).
14. Gayer, R.A. & Jones, J. The Variscan foreland in South Wales. *Proceedings of the Ussher Society*, **7**, 177-179 (1989).
15. Gayer, R.A. The effect of fluid over-pressuring on deformation, mineralisation and gas migration in coal-bearing strata. In: *Contributions to an International Conference on fluid evolution, migration and interaction in rocks*. J. Parnell, A.H. Ruffell, & N.R. Moles (Eds). pp. 186-189. Geofluids '93 Extended Abstracts, Torquay (1993).
16. Gayer, R.A., Cole, J., Frodsham, K., Hartley, A.J., Hillier, B. Miliorizos, M. & White, S. The role of fluids in the evolution of the South Wales Coalfield foreland basin. *Proceedings of the Ussher Society*, **7**, 380-384 (1991).

17. Gayer, R.A., Fowler, R.P. & Davies G.A. Coal rank variations with depth related to major thrust detachments in the South Wales Coalfield; implications for fluid flow and mineralisation. In: *European Coal Geology and Technology.* R.A. Gayer & J. Pesek (Eds). **In Press.** Geological Society, London, Special Publication (1997).

18. Hartley, A.J. & Warr, L.M. Upper Carboniferous basin evolution in SW Britain. *Proceedings of the Ussher Society,* 7, 21-216 (1990).

19. Hartley, A.J. A depositional model for the Mid-Westphalian A to late Westphalian B Coal Measures of South Wales. *Journal of the Geological Society, London,* 150, 1121-1136 (1993).

20. Hathaway, T.M. & Gayer, R.A. Variations in the style of thrust faulting in the South Wales Coalfield and mechanisms of thrust development. *Proceedings of the Ussher Society,* 8, 279-284 (1994).

21. Hillier, B.V. *Seismic studies of deep structure beneath the South Wales coalfield and adjacent areas* [unpublished PhD thesis] University of Wales, Cardiff (1989).

22. Horvàth, Z.A. Study on maturation process of huminitic organic matter by means of high pressure experiments. *Acta Geologica Hungarica,* 26, 137-148 (1983).

23. Jones, J.A. The influence of contemporaneous tectonic activity on Westphalian sedimentation in the South Wales coalfield. In: *The role of tectonics in Devonian and Carboniferous sedimentation in the British Isles.* R.S. Arthurton, P. Gutteridge & S.C. Nolan (Eds). pp. 243-253. Special publication of the Yorkshire Geological Society, Wigley (1989).

24. Jones, J.A. *Sedimentation and tectonics in the eastern part of the South Wales Coalfield* [unpublished PhD thesis] University of Wales, Cardiff (1989).

25. Jones, J.A. A mountain front model for the Variscan deformation of the South Wales coalfield. *Journal of the Geological Society, London,* 148, 881-891 (1991).

26. Kelling, G. Silesian sedimentation and tectonics in the South Wales basin: a brief review. In: *Sedimentation in a synorogenic basin complex, the Upper Carboniferous of north-west Europe.* B. Besly, & G. Kelling (Eds). pp. 38-42. Blackie, London (1988).

27. Littke, R., Büker, C., Lückge, A., Sachsenhofer, R.F. & Welte, D.H. A new evaluation of palaeo-heat flows and eroded thicknesses for the Carboniferous Ruhr basin, western Germany. *International Journal of Coal Geology,* 26, 155-183 (1994).

28. Mukhopadhyay, P.K. Vitrinite Reflectance as Maturity Parameter. Petrographic and Molecular Characterisation and its application to Basin Modelling. In: *Vitrinite Reflectance as a Maturity Parameter, Applications and Limitations,* P.K. Mukhopadhyay and W.G. Dow (Eds). pp. 1-24. American Chemical Society, ACS Symposium series 570, Washington DC (1994).

29. Quick, J.C. Iso-rank variation of vitrinite reflectance and fluorescence intensity. In *Vitrinite Reflectance as a Maturity Parameter, Applications and Limitations,* P.K. Mukhopadhyay and W.G. Dow (Eds). pp. 64-75. American Chemical Society, ACS Symposium series 570, Washington DC (1994).

30. Salih, M.R. & Lisle, R.J. Optical fabrics of vitrinite and their relation to tectonic deformation at Ffos Las, South Wales coalfield. *Annales Tectonicae,* 2, 98-106 (1988).

31. Sibson, R.H. Crustal stress, faulting and fluid flow. In: *Geofluids: Origin, Migration and Evolution of Fluids in Sedimentary Basins.* J. Parnell, (Ed). pp. 69-84. Geological Society, London, Special Publication 78, (1994).

32. Teichmüller, M. & Teichmüller, R. The geological basis for coal formation. In: *Stach's textbook of Coal Geology* (3rd. Edition). E. Stach, M.Th. Mackowsky, M. Teichmüller, G.H. Taylor, D. Chandra & R. Teichmüller (Eds). pp. 5-86. Gebrüder Borntraeger, Berlin (1982).

33. Tissot, B.P. & Welte, D.H. Petroleum formation and Occurrence. 2nd Ed. Springer, Berlin (1984).

34. Veld, H., Fermont, W.J.J., David, P., Pagnier, H.J.M. & Visscher, H. Environmental influence on maturity parameters in Carboniferous coals of The Netherlands. *International Journal of Coal Geology*, **30**, 37-64 (1996).
35. Veld, H., Janssen, N.N.M., Fermont, W.J.J. & Pagnier, H.J.M. Coal facies interpretations and vitrinite reflectance variations in Carboniferous coals from well Limbricht - 1/1a, The Netherlands. *Comptes Rendus XII ICC-P*, **1**, 267-278 (1993).
36. White, S. Palaeo-geothermal profiling across the South Wales Coalfield. *Proceedings of the Ussher Society*, **7**, 368-374 (1991).
37. White, S.C. *The tectono-thermal evolution of the South Wales Coalfield* [unpublished PhD thesis] University of Wales, Cardiff (1992).
38. Wilson, D., Davis, J.R., Smith, M. & Waters, R.A. Structural controls on the Upper Palaeozoic sedimentation in south-east Wales. *Journal of the Geological Society, London*, **145**, 901-914 (1988).
39. Wilks, K.R., Mastalerz, M., Bustin, R.M., & Ross, J.V. The role of shear strain in the graphitization of a high - volatile bituminous and an anthracite coal. *International Journal of Coal Geology*, **22**, 247-277 (1993).

Proc. 30th Int'l Geol. Congr., Vol. 18, Part B, pp. 99-122
Yang Qi (Ed.)
© VSP 1997

COALIFICATION JUMPS, STAGES AND MECHANISM OF HIGH-RANK COALS IN CHINA

QIN YONG and JIANG BO

Department of Geology, China University of Mining and Technology, Xuzhou 221008, Jiangsu, China

Abstract

Based upon the samples collected from more than twenty coal mining districts in China, the coalification jumps, stages and mechanism of high-rank coals were systematically studied in the light of coal petrology and coal chemistry. It was indicated that three stages, namely as meta-exinitic, meta-vitrinitic and meta-inertinitic stages, are undergone in succession during the development of the reflectance of high-rank macerals according to the measured data. A new coalification jump, occurred at about 8% maximum vitrinite reflectance, was found and was considered as one of major indicators in the classification of coalification stages. Six jumps of high-rank coalification were demonstrated in detail, and the Classification Scheme of Coalficaton Stages of High-rank Coals in China was established, in which three stages, i.e., hypo-, meso- and hyper-anthracites, were distinguished respectively at 2%, 4% and 8% maximum vitrinite reflectance. The Makingup, considered as a dynamochemical process and a special mechanism for high-rank coalification, was suggested, and its fundamental patterns and evolution were prelimitarily investigated.

Keywords: China, high-rank coal, coalification jump, coalification stage, makingup

INTRODUCTION

The high-rank coal, called also as anthracite, means the coal with more than 2% oil maximum vitrinite reflectance [2, 10]. The stages and jumps of high-rank coalification are shown as the progressive and sudden changes of the coal properties occur alternatively during the coalification, to which a close attention has been paid by coal scientists.

The previous works demonstrated a few jumps that exist in the process of the coalification from peat to anthracite [22, 27]:

The first jump, occurring at 0.6% mean vitrinite reflectance, 43% volatile matter and 80% carbon, is shown as the raw humic acid and the primary huminite fluorescence in coals disappear and the secondary fluorescence in vitrinite appears [25]. At the jump, the biogenetic gas stage ends and the thermal degraded-hydrocarbon stage starts. That is to say, the coalification develops from brown coal into bituminous coal due to the first jump which corresponds about to the threshold of oil window.

The secondary jump takes place at 1.3% mean vitrinite reflectance, 29% volatile matter and 87% carbon. Near the jump, the fluorescence of coal macerals tends to disappear, the maximum reflectance of some exinite macerals such as sporinite is close or equal to that of vitrinite, the maximum reflectance, refractive index, absorptivity, aromaticity and ring condensation index start to increase dramatically, the optimum coking ability is obtained, and the density and microhardness are decreased to a minimum. The jump is matched with the oil dead line as the oil window ends and the thermal cracking gas begins to being substantially generated.

The third jump exists at about 2.5% mean vitrinite reflectance, 8% volatile matter and 91% carbon. About near 2.0% vitrinite reflectance, the local molecular ordering of coal increases remarkably and the maximum of coalbed methane generation comes as the fatty groups are substantially degraded off from the aromatic condensed ring system of coals.

The fourth jump, occurring at 3.7% mean vitrinite reflectance, 4% volatile matter and 93.5% carbon, is remarked by the reversals of vitrinite refractive index and the aromatic cluster height and the dramatical drop of free radical intensity in coals. The jump tends to the gas-generated dead line.

Moreover, the reversal of minimum vitrinite reflectance at 6.5% maximum vitrinite reflectance was described by Ragot for the first in 1977 [22], and the sudden occurrence of graphitization from coals was found by Oberlin in 1975 [14], using the transmitted electron microscopy. Those implied that, in the process of high-rank coalification, there might be some other jumps which remain to be further demonstrated.

It can be seen from the works above-mentioned that the more attention has been paid to the coalification jumps of low- and medium-rank coals other than that of high-rank coals. As the result, some issues on high-rank coalification remain to be deeply discussed:

Firstly, the course of high-rank coalification ranges from 2% to 11% maximum vitrinite reflectance, much longer than that of low- and medium-rank coals. During the long coalification, though some jumps have been found and studied, the more systematical investigation should be made on the characteristics of these jumps, specially for those with more than 6% maximum vitrinite reflectance, and other jumps, possibly existing in the process of coalification, might remain not to be found. Moreover, some previous conclusions were made mainly in accordance with the artificial coailification experiments. In order to testify the conclusions, the natural coalification should be studied in more detail.

Secondly, the insufficient knowledge on the jumps might result in the poor understanding to high-rank colification stages. In other words, the characteristics of high-rank coalification in the nature could not be relected comprehensively by the previous schemes of coalification stages. For example, the coalification process with more than 4% maximum vitrinite reflectance is not subdivided in the schemes of ASTM and DIN [22], and the 6% maximum vitrinite reflectance is considered only as the latest boundary of coalification stages in the scheme of China, and 5.5% as that in the scheme of the pre-Soviet Union [15, 12].

Thirdly, the essence or mechanism of high-rank coalification remain to be deeply discussed. It has been found at present that, during the middle to late coalification of high-rank coals, the molecular basic structural unit (BSU for short) develops remarkbaly while the evolution of organic element composition and aromaticity is very slow and is even at a standstill and the signal of electron paramagnetic resonance (EPR for short) appears notably again. Those could not be explained according to the traditional theory of aromatization and/or ring condensation.

The high-rank coals occur widespreadly in the major coalfields or basins of China and in all the Paleozoic to Mesozoic coal-bearing formations. There is a complete range of coal ranks in China, ranging from anthracite to graphite. The reserves of high-rank coals make up more than 18% of total coal reserves and are the important part of coal resources in China [6]. Consequently, these are in favorable of a deep study on the characteristics and mechanism of high-rank coalification in China.

Basing upon the favorable condition, the authors studied the stages and jumps of Chinese high-rank coalification in the light of the testing and analyzing data of sample series from bituminous coal to graphite, found a new jump at about 8% maximum vitrinite reflectance in the same time that the previous jumps were systematically examined, and, then, suggested and discussed the makingup, a special mechanism for high-rank coalification.

SAMPLES AND METHODS

The samples for the study of maceral reflectance, amounting to about 200 pieces, were collected from 22 coal mining districts in which the major high-rank coalfields in China locate (Table 1). The maximum vitrinite reflectance of the samples varies from 1.32% to 17.90%, including not only all the range of high-rank coalification but also the part of medium-rank coals as well as graphite. The microscope photometer MPV II made in the German LEITZ Company was used to measure the reflectance under the conditions of oil object with 32X magnification. The measurements for the main individual macerals in each sample are more than 30 counts.

A set of the samples for the analysis of coal structure was chosen from the samples above-mentioned for the reflectance, consisting of the hand-picked vitrain, clarain and graphite, and their fundamental properties were listed in Table 2. The subsamples used to testing the chemical structure were pre-treated by HCl, HF and HNO_3 so as to removing the mineral matters such as carbonate, oxide, silicate and sulfide [25].

The volume, diameter and their distribution of the pores in coals were measured using the Micromeritics Instrument 9310 made in USA under the pressure of 3.5 kPa to 206787 kPa corresponding the minimum pore diameter of 7 nanometers. The samples were 1-2 millimeters in size. According to the China National Standard GB 217-81, the

Table 1. Distribution of the samples used in this paper

location	$R_{o,max}$ (%)	geological era
northern China		
Yuxian	1.83-1.96	the Upper Carboniferous and the Lower Permian
Xinmi	2.37-3.37	the Upper Carboniferous and the Lower Permian
Xingong	4.70-6.25	the Upper Carboniferous and the Lower Permian
Yanlong	5.60-6.45	the Upper Carboniferous and the Lower Permian
Jiaozuo	3.64-5.50	the Upper Carboniferous and the Lower Permian
Jiyuan	2.07-6.65	the Upper Carboniferous and the Lower Permian
Xingtai	17.90	Archeaozoic
Jingxi	5.98-11.23	the Upper Carboniferous and the Lower Jurassic
Jincheng	3.75-4.20	the Lower Permian
Hancheng	1.66-2.06	the Upper Carboniferous and the Lower Permian
northwestern China		
Ruqigou	3.54-3.58	the Lower and Middle Jurassic
southern China		
Liuzhi	1.32-2.06	the Upper Permian
Zhina	2.96-3.93	the Upper Permian
Xinzhong	2.97-5.12	the Lower Carboniferous and the Upper Permian
Meitian	2.00-13.00	the Upper Permian
Shaowu	6.99-15.00	the Upper Jurassic
Yongan	6.19-6.92	the Lower Permian
Tianhushan	7.12-8.40	the Lower Permian
Zhangping	1.82-2.54	the Upper Trassic
Longyan	4.74-4.99	the Lower Permian

apparent and true densities of the samples was measured using the method of water exchange and, then, total porosity was computed.

The proximate analysis of the samples was made in accordance with the China National Standard GB 212-91 and the ultimate analysis with the GB 476-91. Then, the aromaticity and the ring condensation index were computed using the method of statistical constitution analysis [23].

The size of BSUs or aromatic clusters was measured by the X-ray diffractometer D/max-IIB made in the Japan Rigaku Company. The conditions used to measuring were as follows: CuKa target, 40 kV tube current, 30 mA tube voltage, 0.50 SS, 0.50 DS, 0.15 mm RS, step-advanced scanning, 0.15o 2Q step space, 2S inherent time, and 1o/min scanning speed. The corrections such as air diffusion, polarization factor, Compudon diffusion and absorption factor were made before the data related to BSUs was computed.

According to the Oberlin's method [14], the BSUs of the samples were observed and measured by the transmitted electron microscope JEM-200CX made in the Japan JEOL Company under the bright field (BF), dark field (DF) and selected area diffraction (SAD) with 35 mA current and 160 kV voltage. The microphotographs were taken under the magnification of 20,000 to 330,000 times.

The signals of EPR, including free radical intensity (Ng), linewidth (delta H) and spectroscopic splitting factor (g value), were determined using the electron magnetic resonance spectrometer E-109 made in the American Varian Company. The Strong Pitch (1,1-diphenyl-2-picrylhydrazyl, DPPH for short) was used as a standard.

EVOLUTION OF PHYSICAL AND CHEMICAL PROPERTIES OF HIGH-RANK COALS

Development of maceral reflectance

The remarkable achievements on the development of the optical properties of high-rank macerals were reported by some previous reseachists [1, 22, 24, 20]. In this study, the authors not only manifested again some previous works but also gained new lights or knowledge such as the evolution of macrinite reflectance and the pattern of high-rank maceral reflectance.

Evolution of macrinite reflectance
The macrinite exists in the most of Chinese high-rank coals and its polished surface is enough large to meet the need of reflectance measurement. Based on the data measured, it was found that the maximum macrinite reflectance takes a reverse at about 8% maximum vitrinite reflectance other than a sustained increase (Fig. 1a and 1b).

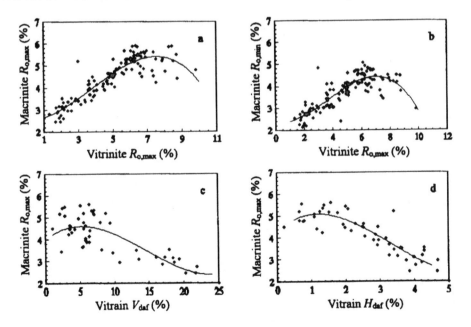

Figure 1. Development of high-rank macrinite reflectances. Ma means macrinite, and V symbols for vitrinite.

In the meantime, the reversal of minimum macrinite reflectance takes place nearly at 7% maximum vitrinite reflectance, which lags behind the reversal of minimum vitrinite reflectance.

A same trend was revealed by comparison with the chemical properties of the vitrains and clarains from the same samples. The reversal of maximum macrinite reflectance corresponds respectively to 1.5-1.0% hydrocarbon content and to 5-2% volatile matter yield (daf basis) after which the reflectance decreases remarkably (Fig.1c and 1d).

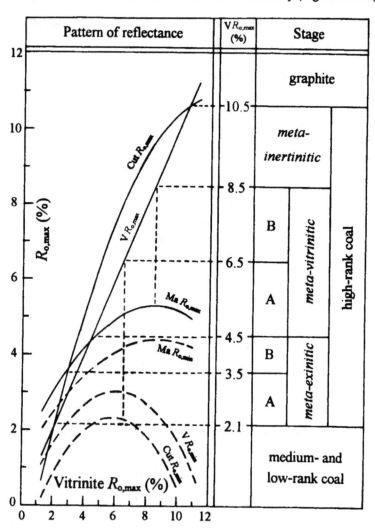

Figure 2. Pattern of high-rank maceral reflectances. V symbols for vitrinite, CUT for cutinite and Ma for macrinite.

Pattern of high-rank maceral reflectance
Based upon a large number of reflectance data measured, the pattern of maceral reflectance of high-rank coals in China was suggested by the authors (Fig.2). The

pattern showed that the reflectance of high-rank macerals develops through three stages.

The first stage, named as meta-exinitic one occurred between 2.1% and 4.5% maximum vitrinite reflectance, is characteristic of the excess of maximum cutinite reflectance over maximum vitrinite reflectance, resulting in the reversal of the interrelation between the two reflectances. The stage could be subdivided into two substages A and B on the basis of the excess of maximum cutinite reflectance over maximum inertinite reflectance about at 3.5% maximum vitrinite reflectance.

The second stage, called as meta-vitrinitic one existed between 4.5% and 8.5% maximum vitrinite reflectance, shows that the maximum reflectance of vitrinite exceeds that of macrinite. As the result, three reversals of the interrelation among the reflectances of three maceral groups throughout the coalification process are thoroughly accomplished. Two substages in the stage could be distinguished with reference to the reversal of minimum vitrinite reflectance at about 6% maximum vitrinite reflectance.

The third stage, named as meta-inertinitic one, is characterized by the notable reversal of maximum macrinite reflectance and ranged from 8.5% to 10.5% maximum vitrinite reflectance. Some previous reseachists considered the stage as the semigraphite one or even as graphite one [11, 21]. However, the result derived from the microscopic observation showed that the microtexture or morphological characters of the macerals was well preserved, the macerals could be clearly distinguished and any trace of graphitization was not found even in the samples with 8.0-9.7% maximum vitrinite reflectance. In other words, these samples have the typical characters of coals, which has been demonstrated by chemical structure analysis (see after).

At 10.5% maximum vitrinite maximum, the maximum reflectance of cutinite tended to be equal to that of vitrinite once again, indicating the end of coalification or the start of graphitization. The maximum reflectance of crystalline graphite measured in the study was 17.90 percent, generally corresponding to the end of graphitization [19].

It is known that the maceral reflectance mirrors the chemical structure and other physical properties of coals in some degree as an important physical property. For this reason, the authors made a further investigation on high-rank coal structure so as to bring much deeply to light the characteristics and mechanism of high-rank coalification.

Evolution of micropore structure

Coal is a substance with highly developed system of micropores. The micropores in coals has been previously studied well by many reseachists, and the scheme of coal pore classification with the decimal system, suggested by Xoxot in 1961 [26], has been popularly applied to the coal industry, and the schemes respectively suggested by Dubinin in 1960 and by Gan in 1972 can be seen commonly in the publication on coal chemistry and physics [8]. However, the distribution of the micropores in high-rank coals is slightly different from that in bituminous coals, which was specially studied by the authors in 1995. As the result, the classification of pore structure of high-rank coals was suggested by the authors as follow: macro-pore with the diameter of more than 450 nm, meso-pore with that of 450 nm to 50 nm, interim-pore with that of 50 nm to 15 nm, micro-pore with that of 15 nm to 7 nm and submicro-pore with that of less than 7 nm [16]. Next, the distribution of pore in high-rank coals of China would be examined according to this classification.

The figure 3 shows the development of pore volume and porosity of high-rank coals in China. It can be seen that, with maximum vitrinite reflectance increasing, the total pore volume, the fractional pore volume and total porosity synchronize to develop, and reversed respectively at 4.5%, 8.5% and 11% maximum vitrinite reflectance. Based on those, all the coalification process of the pore structure in high-rank coals could be distinguished into three stages, i.e., increasing, decreasing and re-increasing ones, which corresponds to the stages of maceral reflectance. Of course, so is the development of specific pore surface area.

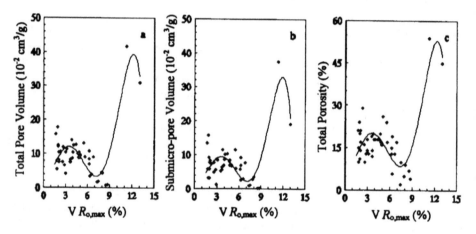

Figure 3. Evolution of micropore volumes and porosity of high-rank coals.

Compared Fig.3a with Fig.3b, it can be known that the volume of the pore with less than 7 nm of diameter is predominant over total pore volume no matter how the stage is. In other words, The volume of this pore makes up a main part of total pore volume. Generally, the macro-pore is derived from the micro-fissure or micro-cleats remained in coal samples, the meso-pore from the inner- and outside-pore of coal macerals, and the micro-pore and submicro-pore from the inter-BSU pore ("cavity" existed between the BSUs) [28]. Thus, the most of the pores in high-rank coals could be composed of the inter-BSU pore.

This study showed that the size and ordering of the BSUs in coals develop highly once the maximum vitrinite reflectance is over more than 8% (see after). Therefore, the volume of the inter-BSU pore should seem to decrease theoretically with maximum vitrinite reflectance so that total pore volume decrease, but it was contrary in fact. Thecontradictory against the traditional theory might be associated closely with the mechanism of high-rank coalification (see after).

Evolution of macro-molecular basic structural units (BSU)

The fundamental characteristics of the BSUs in coals can be indicated by the parameters such as the stacked height, the diameter and the layer space of BSU in which the development of the height and diameter is most remarkable. The data from the X-ray diffraction (XRD) analysis showed that the reversals of the BSU height and the "stair-like" jumps of the BSU diameter exist in the process of high-rank coalification (Fig.4).

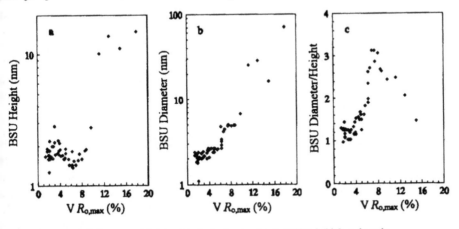

Figure 4. Evolution of diameter and height of the basic structural units (BSUs) in high-rank coals.

The reversals of the height take place respectively at two jumps, with one ranging from 3.5% to 4.0% maximum vitrinite reflectance corresponding to 93% carbon and 3.0% hydrogen, and with another locating near 6.0% maximum vitrinite reflectance corresponding to 96% carbon and 2.0% hydrogen, which presents three stages of BSU development (Fig.4a). The height increases progressively by 0.2-0.4 nm in the first stage, decreases gradually by 0.4-0.5 nm in the second stage, and increases once again in the third stage so that a reversal occurs at about 6.0% maximum vitrinite reflectance. However, the rate of the height development in the third stage is not constant. The height increases only by about 0.4 nm when maximum vitrinite reflectance ranges from 6.0% to 8.5%, but increases remarkably once the maximum vitrinite reflectance is more than 8.5%.

It should be noted that, when maximum reflectance exceeds 11%, the development of BSU height slows down notably. This indicates on the one hand the end of coalification, and, on the other hand, the abrupt occurrence of the transformation from amorphous coals to crystalline graphite.

As the maximum vitrinite reflectance increases, the BSU diameter of high-rank coals tends to increase jumpily, which indicates a stair-like pattern with flat steps. The first step with the BSU height of 2.1 nm extends from 2.0% to 4.0% maximum vitrinite reflectance, and the second step with the height of 2.6 nm from 4.0% to 6.0% maximum vitrinite reflectance. As the maximum vitrinite reflectance rises from 6.0% to 7.0%, the BSU height steps up to 4.8-5.0 nm   until 8.5% maximum vitrinite reflectance or so, by which the third step is shown (Fig.4b). After this, the height increases jumpily again.

Two dividing lines among three steps of the BSU diameter correspond respectively
two reversing lines of the BSU height. It should be pointed out that the ratio of th
diameter to the height takes a reversal at 8.0% maximum vitrinite reflectance (Fig.4c
This provided more information for the jump of BSUs during high-rank coalification.

In general, coals are considered as a matter with short ordering but without lon
ordering. As the degree of coalification increases, the ordering degree of the BS
arrangement in three-dimensional space would be strengthened and the shape of th
BSU would be changed so that the short ordering structure of coals could b
transformed finally into the long ordering structure of graphite. Using the transmitte
electron microscopy, not only this evolution can be much deeply revealed but also mor
information on coal structure may be inquired.

During the initial coalification of high-rank coals (2.0-4.0% maximum vitrinit
reflectance), the misorientation between two adjacent BSUs is about 30° statistical
with the maximum of 60° (Fig.5c), being similar to the turbostratic-layer patter
suggested by Franklin in 1951 [9]. However, the Franklin's pattern means the stacke
style of aromatic layers inside BSU but the turbostratic pattern in this paper refers to th

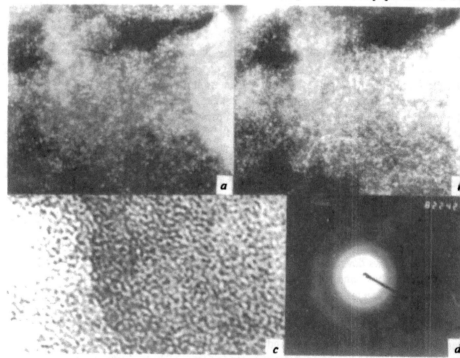

Figure 5. TEM images of the Jiaozuo vitrain with 3.65% maximum vitrinite reflectance. (a) and (b), 002 dark
field images, showing the azimutal misorientation of BSUs represented by the bright dots, × 294,900, objectiv
aperture moved 90° along 002 diffraction ring; (c), bright-field image, indicating the arrangement of BSU
represented by black dots, × 804,200; (d), SAD pattern.

rranged style among individual BSUs, and the latter is different from the former in the magnification of observation. Compared with two 002 dark-field images with the mis-azimuth of 90° along 002 diffraction ring, it could be seen that the close-packed degree nd brightness of the BSU dots show hardly a difference, and the dots occur dispersively o that the porphyritic lumps, which are generated by the close aggregation of BSU dots, ould be seldom found (Fig.5a and 5b). The SAD pattern showed that the 002 and 10 ings diffused much greatly and the 11 ring could be only seen faintly (Fig.5d). Those ndicate that a low degree of BSU ordering remains still in the stage.

During the medium coalification of high-rank coals (4.0-8.0% maximum vitrinite reflectance), the turbostratic pattern of BSUs disappears almost and the statistical misorientation between two adjacent BSUs is 15° or so (Fig.6c). The difference of the close-packed degree and brightness of the BSU dots exist in two 002 dark-field images with the mis-azimuth of 90°, and the porphyritic lumps of BSUs could be clearly seen Fig.6a and 6b). As the maximum vitrinite reflectance increases, the diameter and compactness of the lumps increase, indicating that local molecular ordering (LMO) becomes strong. In the SAD pattern, the 11 ring can be clearly seen and the 10 ring becomes bright and narrow, which implies the remarkable increase of the ordering among the BSUs and inside individual BSUs (Fig.6d).

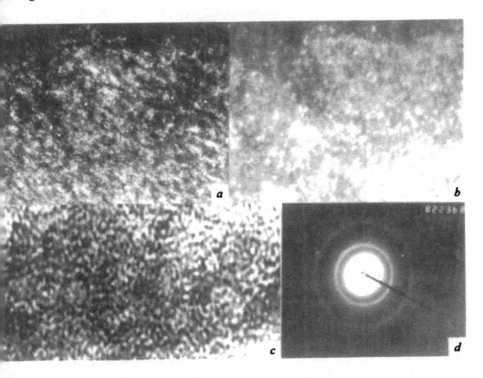

Figure 6. TEM images of the Jiyuan vitrain with 6.23% maximum vitrinite reflectance. (a) and (b), 002 dark-field images, showing the amimutal misorientation of BSUs, × 311,200, objective aperture moved 90° along 002 diffraction ring; (c), bright-field image, indicating the arrangement of BSUs represented by black dots, × 848,600; (d), SAD pattern.

During the late coalification of high-rank coals (more than 8% maximum vitrinit
reflectance), the BSUs are substantially aggregated into the porphyritic lumps, and th
misorientation of BSUs among and inside the lumps tends to become 0° (Fig.7c
indicating the high ordering of BSUs in the stage. However, the curvature of the BSU
in this stage seems to be larger than that in the medium stage (Fig.7a and 7b), whic
might be derived from the makingup of BSUs (see after). The SAD pattern in this stag
is similar to the pattern in the medium stage with the exception of the more narro\
diffraction rings (Fig.7d). Those indicate sufficiently that, though the coal rank of th
samples in this stage is close to graphite, their BSUs are typically   characteristic of co;
other than graphite.

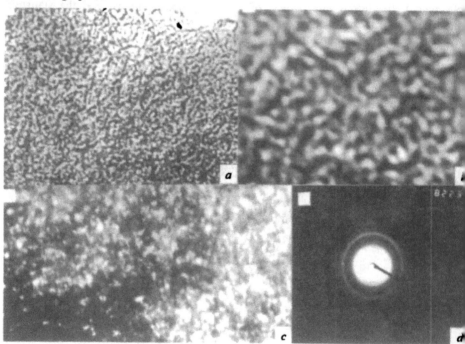

Figure 7. TEM images of the Jingxi clarain with 9.69% maximum vitrinite reflectance. (a) and (b), bright-fiel
images, showing the false curvature of BSUs, × 294,900(a) and +804200; (c), dark-field image, indicating th
remarkable BSU lumps represented by bright dots, × 804,200; (d), SAD pattern.

The transformation from coal to graphite takes place abruptly. Once the graphite stag
comes, the BSUs were rapidly made up, and the diameter of BSU layers increases to :
high degree so that the scaly graphite that can be visible even to the naked eye is forme
(Fig.8). As for the SAD pattern, the diffraction rings are disintegrated into the dotte
rings, and the 100, 101, 110, 112 and 004 rings can be clearly seen (Fig.8d). However
there exist not only the flat BSU clusters with very large diameter, but also a larg
number of fold and poorly continuous clusters in the initial period of graphitization du
to the great difference of the BSU ordering (Fig.8a, 8b and 8c). As the maximun
reflectance continues increasing, the BSU clusters become flat and straight, the BSl
diameter increases dramatically, and the typical monocrystals of graphite as well as it
hexagonal diffraction dots appear.

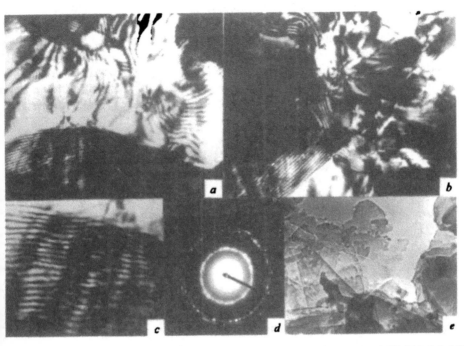

Figure 8. TEM images of the Jingxi graphite with 11.27% maximum reflectance. (a) and (b), 002 dark-field images, indicating the coexistence of fold, micro-columnar and flat graphites, × 140,940; (c), 002 dark-field image, showing the micro-columnar graphite, × 289,600; (d), SAD pattern, showing the dotted diffraction rings; (d), bright-field image, indicting the scaly graphite.

Based upon the results above-mentioned, a model of the BSU evolution from high-rank coal to graphite was suggested by the authors (Fig.9). As the coal rank rises progressively during the high-rank coalification, the BSU ordering is gradually enhanced, the misorientation between two adjacent BSUs becomes small, and the molecular ordering domains develop from isolated BSUs, through loose strawberry-like lumps, to compact stripe-like lumps. The evolution stages of the BSUs are roughly consistent with those of the maceral reflectances and pore properties above-mentioned. The graphitization occurs abruptly. The BSUs could be integrated together end by end so that the diameters of BSU clusters become large rapidly while the fold graphite is transformed dramatically into flat and straight graphite. The boundary between high-rank coal and graphite is very clearly shown under transmitted electron microscope.

The organic matter in coal is mainly composed of five elements, i.e., C, H, O, N and S, whose contents and characteristics of bonding in macromolecular network affect directly the chemical structure of coals. For this reason, the study of the statistical constitutions is helpful to the understanding of the characteristics and mechanism of high-rank coals.

As the maximum vitrinite reflectance increases, the content of the hydrogen in high-rank coals, with very little divergence, decreases regularly and constantly, and so it is even in the late coalification of high-rank coals (Fig.10a). Even though the content of

the carbon increases with reflectance increasing, its divergence would become mo.
remarkable after more than 4% maximum vitrinite reflectance (Fig.10b). Those sho
that the content of the hydrogen maybe be considered as a more suitable indicator
the coal rank during the high-rank coalification than that of the carbon. It could b
found in the meantime that the contents of the hydrogen and carbon reduce much fa
before less than 4% maximum vitrinite reflectance than after more than 8% maximu
vitrinite reflectance during which the content of the carbon remained almost constant.

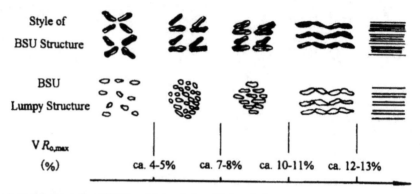

Figure 9. Model of evolution of BSUs and their ordering of high-rank coals.

Evolution of organic elements and statistical constitutions

As for the statistical constitutions, the aromaticity increased dramatically when th
maximum vitrinite reflectance was less than 4%, then remained almost consta
(Fig.10c). Similarly, the development of ring condensation index $2(R-1)/C$ becam
suddenly very slow (Fig.10d). However, it should be noted that the BSUs developed ve
fast once the maximum vitrinite reflectance was more than 8 percent, implying tha
there exists possibly some special mechanism of coalification different from
aromatization and ring condensation in the medium to late stage of high-ran
coalification (see after).

The figure 10 shows roughly also that three stages exist during the development
organic elements and statistical constitutions of high-rank coals, i.e., the fast developin
one with less than 4% maximum vitrinite reflectance, the transitional one from 4%
8% maximum vitrinite reflectance and the almost stagnated one with more than 8
maximum vitrinite reflectance.

Evolution of EPR signals

Coal is a kind of the matter with paramagnetism, and a great deal of the usef
information about the amount and chemical micro-environment of the unpaired
electrons in coal can be obtained from the signals of electron paramagnetic resonanc
(EPR) which is thus used widespreadly as a effective tool to study coal structure. Th
EPR spectrum of coal is usually expounded as three basic parameters including fre

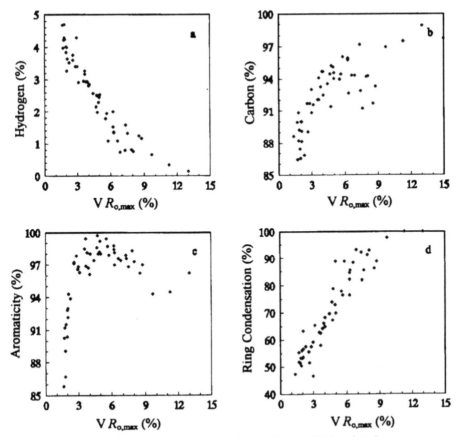

Figure 10. Evolution of organic element contents and statistical constitutions of high-rank coals.

radical intensity (Ng), linewidth (delta H) and spectroscopic splitting factor (g value). The Ng represents total energy absorbed by coal samples under the condition of resonance, is directly proportional to the concentration of the unpaired electrons in samples and related to the aromaticity of the organic matters in coal. The g value is associated not only with the aromaticity of coal but also with the atom O or S existed near the paramagnetic centers, and tends to decrease during the early and medium stages of coalification and to increase dramatically at the end of coalification [18]. The delta H is affected by the interaction of electron self-spin with its chemical micro-environment, and becomes wide as the amount of the atom H, N or S near the paramagnetic centers in coal increases.

As shown in Fig.11, the EPR signals of high-rank coals develop notably in three stages. In the first stage, the Ng of the samples raises progressively and the delta H reduces slowly, but the two drop suddenly about at 4% maximum vitrinite reflectance. In the second stage, the Ng is very low and remains almost constant, the delta H increases slowly, and, then, the two increases jumpily during 6-8% maximum vitrinite reflectance, with the delta H rose by times. In the third stage, the Ng continues still increasing, but the delta H drops dramatically due to the reversal in the initial stage at 8% maximum

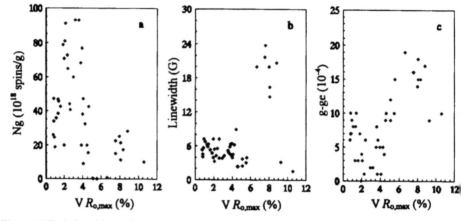

Figure 11. Evolution of free radical intensity, linewidth and g value of high-rank coals.

vitrinite reflectance. The g value reduces slowly when the maximum vitrinite reflectance is less than 4%, takesa reversal at 4% maximum vitrinite reflectance and raises sharply when the reflectance is more than 6%.

It is worth notice that the delta H reverses at 2% maximum vitrinite reflectance and the Ng and g value reverse once again about at 10% maximum vitrinite reflectance. Those provided more information of coal structure for determining the start and end of high rank coalification.

DISCUSSION

Jumps of high-rank coalification

The results revealed that there are six sudden changes of the physical and chemical properties of coals in the process of the coalification from high-rank coals to graphite so that six jumps are shown in various way and in various degree.

The first jump
It appears about at 2.0% maximum vitrinite reflectance corresponding to 89% carbon content of vitrains and clarains.

The jump is mainly characteristic of the exceeding of maximum cutinite reflectance over maximum vitrinite reflectance, the sudden enhance of BSU diameter and the reversal of the linewidth. According to the Oberlin [13,14], the BSU diameter and LMC in vitrinite group increase suddenly at 1.8-2.0% mean vitrinite reflectance (corresponding to 2.0-2.2% maximum vitrinite reflectance), which was considered as the results derived from the removal of the most of fatty groups and oxygen-bearing compounds [22]. The occurrence of this jump indicates the start of high-rank coalification or the end of medium-rank coalification.

The second jump
The jump occurs about at 3.0% maximum vitrinite reflectance corresponding to 91% carbon and 10% volatile matter and 3.5% hydrogen. Near the jump, the maximum cutinite reflectance exceeds over the maximum inertinite reflectance, and the BSU lumps in coal start to appear.

The third jump
The jump occurs among 4.0-4.5% maximum vitrinite reflectances about corresponding to 93% carbon, 6% volatile matter and 3.1% hydrogen.

As for the physical properties of coals, the maximum vitrinite reflectance is over the maximum inertinite (macrinite) reflectance so that the overall reversal of the correlation among maximum maceral reflectances is accomplished, and, in the meantime, the pore volume and porosity of vitrains and clarains reach a maximum so that the pore surface area is at a maximum. So far as the chemical property goes, the development of the aromaticity becomes suddenly slow, and the ring condensation index keeps up a substantial increase; the BSU height attains a maximum, the BSU diameter takes a jump, the turbostratic pattern of BSUs disappears and the makingup of individual BSUs takes commonly place; the Ng and delta H drop down suddenly, and the g value starts to increase slowly.

The fourth jump
It exists about at 6.0-6.5% maximum vitrinite reflectance corresponding to 96% carbon, 3% volatile matter and 1.5% hydrogen. The jump is characteristic of the reversals of minimum vitrinite and cutinite reflectances, the minimum of BSU height, and the jumpy increase of BSU diameter, the dramatic increase of ring condensation index, delta H and g value. However, the pore structure has no correspondence at the jump.

The fifth jump
The jump occurs about at 8.0-8.5% maximum vitrinite reflectance corresponding to 1.0% hydrogen. There is no indicator of the jump in both volatile matter yield and carbon content because the former has only a very little decrease and the latter has a very great dispersion.

Physically, the reversal of maximum macrinite reflectance occurs, the pore volume and porosity takes a minimum so that the they reverse once again. Chemically, the development of organic element contents and statistical constitution tends to stagnate, the diameter and height of BSUs increase notably, the ratio of diameter to height reverses, the BSU lumps with stripe structure appear commonly so that the LMO develops very greatly, and the delta H reaches a maximum and, then, takes a reversal.

The sixth jump
The jump exists at 10.5% maximum vitrinite reflectance corresponding to 97-98% carbon and less than 0.5% hydrogen.

The jump is characteristic of the abrupt changes of physical and chemical properties.

For example, the maximum reflectance of cutinite tends to become consistent with that of vitrinite, the porosity, pore volume, Ng and g value reverse and then decrease dramatically once again, the BSU diameter increases dramatically, and the local molecular ordering is replaced by the overall molecular ordering so that the flat and straight carbon network with three-dimension ordering of crystals is formed. Those indicate the end of coalification and the start of graphitization.

Stages of high-rank coalification

The coalification jump means the relatively sudden changes of substance properties, and the combination of some jumps constitutes the coalification stages. Therefore, the position at which the jump occurs can be inevitably considered as the most suitable and natural boundary for the classification of coalification stages. Compared with the showed characteristics above-mentioned, it can be found that the physical and chemical parameters were displayed at every jump in various degrees. The second and fourth jumps were expressed only by a part of parameters, but the first, third and fifth jumps were manifested almost by all the parameters. As for the sixth jump, the sudden changes of BSUs was notably presented, but the organic elements and statistical constitution did not respond to this jump, which reflects a evitable result derived form the fact that the chemical structure in the latest stage of high-rank coalification is constituted almost by pure carbon network. For this reason, the authors divides the process of Chinese high-rank coalification into three stages, named respectively as hypo-anthracitic, meso-anthracitic and hyper-anthracitic ones, in which the former two stages could be subdivided respectively into two substages A and B in accordance with the second and fourth jumps (Table 2).

As concerns the dividing boundary at 4% maximum vitrinite reflectance, the attention was paid in some degree by some reseachists. However, the insufficient knowledge for the fifth jump at 8.5% maximum vitrinite reflectance had been inquired for a long term. In this paper, the authors suggested formally the fifth jump for first time, demonstrated systematically its occurrence, and used the jump as a dividing boundary of coalification stages for first time.

The end of coalification or the start of graphitization was previously discussed, some suggestions such as 11.0%, 10.0% and 8.5% maximum vitrinite reflectance were put forward but no agreeable opinion has been achieved up to now [21, 22, 11]. The study on the sixth jump in this paper shows that the transformation from short ordering coal to long ordering graphite occurs abruptly among 10-11% maximum reflectances during which the sudden changes of chemical and pore structures of coals exist notably. So, the authors suggest 10.5% maximum reflectance as the end of coalification.

Makingup: a special mechanism for high-rank coalification

As stated before, the organic elements and statistical constitutions develop very slowly or even hardly during the medium and late stages of high-rank coalification. In the meantime, the BSUs develop fast and the EPR signals (specially delta H) increase dramatically once again. Those could not be reasonably explained in accordance with the traditional theory such as aromatization and ring condensation. In other words, some special mechanism might exist during those stages of coalification.

Fundamental patterns and their development of makingup
Basing upon the observation under transmitted electron microscope, the authors have suggested the model of BSU evolution and its ordering (Fig.9). It could be seen from the model that the development of BSUs might be realized through three fundamental ways or patterns as follows:

(1) Linkup: the BSUs are linked up in the end by end (Fig.12a), resulting in increasing the diameter of BSUs or ordering domains (OD) by which the new BSUs are formed once the vestiges derived from the linkup disappear.

(2) Pileup: new BSUs are make up in the face by face (Fig.12b), resulting in increasing the height of ordering domains, but the vestiges could be seen still until the stage of graphitization, that is to say, the individual BSUs could be distinguished yet.

(3) Makeup: it is a complex pattern with the characteristics of both linkup and pileup, and includes two styles, i.e., the simple and complex makeups (Fig.12c). The simple makeup might be realized through incomplete linkup and pileup of BSUs, the individual BSUs could be distinguished when the vestige of makeup could be faintly seen, and the false bending of BSUs is formed once the vestige disappears. The complex makeup might be accomplished through the full composite of both linkup and pileup, is expressed as a BSU lump with stripe structure, and might be a kind of higher-ranking style of makeup.

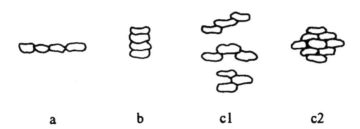

a	b	c1	c2

Figure 12. Fundamental patterns of BSU makingup. (a), linkup; (b), pileup; (c), makeup, including simple makeup (c1) and complex makeup (c2).

The fundamental patterns of makingup develop as the coal rank increases. When the maximum vitrinite reflectance is less than 4.0%, the BSUs exist in a dispersed and isolated form, and the makingup is seldom seen (Fig.5a and 5b). When the maximum vitrinite reflectance reaches 4.0-6.0%, the makingup occurs commonly in coals, might be shown mainly as the linkup in the initial period of the stage so that the BSU diameter increases and as both linkup and simple makeup in the late period of stage (Fig.6a and 6b). When the coal rank reaches 7.0-8.0 % maximum vitrinite reflectance or so, not only the complex makeup develops highly so that the BSU lumps of which the stripe

Table 2. Proposal for the Scheme of Classification of high-rank coalification stages in China.

parameter of coalification	medium-rank coal (bituminous)	hypo- A	hypo- B	meso- A	meso- B	hyper-	graphite
major							
$R_{o,max}$(%)	2.0	3.0	4.0	6.0		8.0	10.5
H_{ad}(%)			3.0	2.0		1.0	<0.5
supplement							
M_{ad}(%)	minimum			increasing		maximum	decreasing
C_{ad} (%)	89	91	93	96.5		?	>98
V_{ad} (%)		10	6	3		?	
reference							
L_c	increasing			maximum	minimum	increasing	
V_0	increasing			maximum		minimum	maximum
N_g	increasing			maximum	minimum	increasing	

structure is dominant are formed, but also the simple makeup exists widespreadly, which leads to the false appearance that the individual BSUs in this stage are more curved than those in earlier stage (Fig.7a, 7b and 7c). During the graphitization, the makingup becomes the sole way by which the LMOs increase. In the initial period of graphitization, the makingup might be characteristic of the coexsitence of both pileup and complete makeups, and result in the generation of the minrocolnma and wrinkled graphites described by Oberlin et al. in 1984 (Fig.8a, 8b and 8c) [13]. However, this period is very short, and seems to continue only for the coalification range of about 2.0% maximum reflectance so that the makingup is finished, and, then, the minrocolnma graphites are linked each other, and the wrinkled graphite disappears and the graphite with typical flat and straight layers is formed.

Probable mechanism of makingup
Based upon the state above-mentioned, the makingup is a kind of the action through which the ordering domains in coal increase, and originated possibly from the manifold reasons. However, no matter how many the reasons are, they must overcome the defects or gaps around and among the BSUs. These defects or gaps might be resulted from non-aromatic groups, heteroatom-bearing bridge bonds and, even, micro-pores among BSUs. Though the mechanism of makingup could not be satisfactorily explained now in accordance with the present data of coal structure, it could be preliminarily elucidated based upon some scientific information.

Firstly, the strong EPR signals in the late stage of high-rank coalification indicates that homologisation must happen in the aromatic structure with condensed rings and the organic free radicals must have been formed. The homologisation should occur at the defects of chemical structures so as to reduce the free energy of the structure and to benefit the transformation toward more stable graphite-like structure. As evidenced by the dramatic increase of the delta H, the defects might locate near the heteroatom O, N and S in heterocyclic structures, which, together with the larger aromatic system, might lead to increasing the g value. The cause and effect correlation of EPR signals to chemical structures of coals was elaborated by Retcofsky in 1982 [18]. In the light of those, it could be reasonably considered that the homolytic cleavage of chemical bonds might be one of the most probable ways by which the heteroatoms in coals are released from the chemical structure in the medium to late stages of high-rank coalification.

Secondly, the process above-mentioned might be realized in two procedures with various mechanisms. In the first procedure, the homologisation might take place and be originated from the higher temperature of paleogeothermal field because the aromatic structures with heteroatomic ring lose their stability to transform into the more stable carbon networks when the temperature rises to certain degree. In the second procedure, the makingup might occur and be greatly promoted mainly under the pressure, special tectonic stress, which was sufficiently manifested by the previous artificial coalification experiments [3, 4, 5, 7, 17]. In China, the occurrence of high-rank coals with more than 6.0% maximum vitrinite reflectance is almost associated with the large-scale late Mesozoic granitic bodies, and the magmatic intrusion provides not only the essential thermal energy but also the powerful tectonic stress for the makingup. The stress might make the aromatic free radicals or the condensed-ring macromolecules close to each other so that the "chemical attraction" between them could interact, which promotes the development of makingup. As for the interrelation between the thermal-deduced homologisation and the stress-deduced makingup, the former is a prerequisite for the realization of the latter, the latter is an end-result for the products of the former, and the more stable macromolecular structures might be the product made comprehensively by the two. As the result, the makingup might be neither a purely chemical function nor a purely physical function, and it should be a dynamochemical process. The stress might play a very important role in the process, because the activation energy needed for the graphitization would be up to 260 kcal/mol that could not be provided even though the temperature up to 700 ℃ last for the whole life interval of the earth [3].

Thirdly, the essence of makingup might be different from that of ring condensation though some similarities between them exist, and, thus, they should represent two mechanisms respectively acted in various stages of coalification. In the first place, the makingup might promote not only the linkup of aromatic networks in "horizontal axis" but also the pileup of the networks in "vertical axis," but the ring condensation can only lead to the bonding in "horizontal axis". In the second place, the ring condensation is resulted mainly from the enhance of coalification temperature, but the makingup might act through the effect of stress as well as the enhance of temperature. In the third place, the makingup might refer hardly to the constitution of organic elements and lead seldom to the formation of new compounds with small molecular weight such as H_2O, but the ring condensation is to the contrary. Consequently, the makingup could result in the increases of both diameter and height of BSUs, and the ring condensation could lead only to the increase of the diameter.

Based upon the chemical mechanism above-mentioned, some characteristics of the coal structures observed could be reasonably explained. Though three fundamental patterns of makingup could be seen in the medium stage of high-rank coalification, the makeup and pileup of the makingup become predominant after the fifth jump. This might lead not only to increasing simultaneously the BSU diameter and height, but also to the falsely curved BSUs (derived from the simple makeup), and the striped BSU lumps and minrocolnma graphites (derived from the complex makingup) (Fig.8). The makingup might correspond to a reaction of free radical type and refer hardly to the loss or enrichment of organic elements or the formation of the compounds with small molecular weight, which lead inevitably to the result that the organic element composition and the statistical constitution do not change but the BSUs develop highly. The makingup might destroy the original three-dimension arrangement of BSUs, and result in the reversals of the porosity respectively at 4.0% and 8.0% maximum vitrinite reflectances because the pores in coals might be made up mainly by the inter-BSU pores of coals.

CONCLUSIONS

Based upon the testing and analyzing data of the samples from more than 20 typical mining districts in China, the authors acquired some new knowledge on the evolution of the structure of high-rank coals as follow:

(1) There exist six jumps of maceral reflectance development in the process of high-rank coalification, based upon which three major stages of high-rank maceral reflectances can be distinguished, i.e., meta-exinitic, meta-vitrinitic and meta-inertinitic ones. Two substages, A and B ones, are included in each of the former two stages.

(2) There occur six jumps of coal pore structure and chemical structures in the process of high-rank coalification. Among them, the third jump about at 4.0% maximum vitrinite reflectance and the fifth jump at 8.0% maximum vitrinite reflectance are displayed in the all-around way of physical and chemical properties, and, therefore, the two jumps could be considered as the most important divides for the stages of high-rank coalification. From this, the authors classified the process of Chinese high-rank coalification into three stages (hypo-, meso- and hyper-anthracites) and five substages, and pointed out that the high-rank coalification starts at 2.0% maximum vitrinite reflectance and ends at 10.5% maximum reflectance.

(3) Based upon the practical observations, the makingup, a new concept, was suggested and preliminarily demonstrated, and regarded as a special mechanism for high-rank coalification. It was found that the makingup includes three fundamental patterns (linkup, makeup and pileup), acts commonly in the meso- and hyper-anthracites, and becomes a main mechanism of BSU evolution in the hyper-anthracite. It was suggested that the makingup could be realized possibly through the process of the deploymerization-makingup, should be a dynamochemical function, and be remarkably different from the ring condensation in acting way, factor and product. In other words, the makingup and the ring condensation might represent two different mechanisms that respectively act in different stages of coalification.

Acknowledgements

The study in this paper was supported by National Natural Foundation of China (Project 49472125). The authors are grateful to Prof. Yang Qi and Prof. W. J. J. Fermont for scientific and linguistic revisions of the manuscript. We thank greatly the VSP staffs for the publication of the paper.

REFERENCES

1. B. Alpern and M. J. Lemos de Sousa. Sur le pouvoir reflecteur de la vitrinite et de la fusinite des houilles, *C.r.Acad. Sci. Paris* **271**, 956-959 (1970).

2. B. Alpern, M.J. Lemos de Sousa and D. Flores. A progress report on the Alpern Coal Classification, *Int. J. Coal Geol.* **13:1-4**, 1-19 (1989).

3. M. Bonijoly, M. Oberlin and A. Oberlin. A possible mechanism for natural graphite formation, *Int. J. Coal Geol.* **1:3**, 283-312 (1982).

4. R. M. Bustin, J. V. Ross and Ian Moffat. Vitrinite anisotropy under differential stress and high confining pressure and temperature: preliminary observation, *Int. J. Coal Geol.* **6:4**, 343-351 (1986).

5. R. M. Bustin, J. V. Ross and J.-N. Rouzaud. Mechanism of graphite formation from kerogen: experimental evidence, *Int. J. Coal Geol.* **28:1**, 1-36 (1995)

6. China Administration of Coal Geology. *Map of Coal Mark in China* (unpublished report). Zhuozhou (1992).

7. C. F. K. Diessel, R. W. Brother and P. M. Black. Coalification and graphitization in high pressure schists in New Caledonia, *Contrib. Miner. Petrol.* **68:1**, 63-78 (1978).

8. M. A. Elliott. *Chemistry of Coal Utilization, second supplementary volume.* John Willey & Sons Inc., London (1981).

9. R.F. Franklin. Crystallite growth in graphitizing and non-graphitizing carbons, *Proc. R. Soc., London,* **209:2**, 196-218 (1951).

10. ISO. *Methods for the petrographic analysis of bituminous coal and anthracite: part 1, Defination of terms relating to the petrographic analysis of bituminous coal and anthracite.* ISO/DIS 7401/01 (1982).

11. Li Shengling and Feng Caixia. Coalification mechanism of the Permo-Carboniferous coals in the western Beijing coalfield of China (in Chinese). *Adv. Coal Sci. Tech.* **1**, 17-23 (1991).

12. K.B. Miloruov. *Handbook of Coal Geologists* (in Chinese). China Geology Press, Beijing (1986).

13. A. Oberlin and M. Oberlin. Graphitizability of carbonaceous materials as studied by TEM and X-ray diffraction, *J. Microscopy,* **132:3**, 353-363 (1983).

14. A. Oberlin and G. Terriere. Graphitization of anthracites by high resolution electron microscopy, *Carbon,* **13:4**, 367-376 (1975).

15. Qin Yong. *Supplement to Coal Geology* (unpublished teaching material in Chinese). China University of Mining and Technology, Xuzhou (1987).

16. Qin Yong, Xu Zhiwei and Zhang Jing. Natural classification of the high-rank coal pore structure and its application, *J. China Coal Soc.* **20:3**, 266-271 (1995).

17. Qu Xinwu. Correlation of coal structure to coalification mechanism (in Chinese), *Coal Geol. and Explor.* **3**, 20-27 (1980).

18. H. L. Retcofsky. Magnetic resonance studies of coals. In: *Coal Science 2.* L. G. Martin et al. (Eds). pp.43-83. Academic Press, New York (1982).

19. Shang Junren. *Mineragraphy* (in Chinese). China Geological Press, Beijing (1987).

20. G.C. Smith and A.C. Cook. Coalification paths of exinite, vitrinite and inertinite, *Fuel,* **59:9**, 641-646 (1980).

21. M. Teichmuller. Coalification. In: *Low Temperature Metamorphism.* M. Frey (Ed). pp.114-161. Blackie,

New York (1987).

22. M. Teichmuller and R. Teichmuller. Fundamental of coal petrology. In: *Stach's Textbook of Coal Petrology*. E. Stach et al. (Eds), pp.5-86. Gebruder Borntraeger, Berlin (1982).

23. D. W. van Krevelen. *Coal: Typology-Chemistry-Physics-Constitution*. Elsevier Scientific Publishing Company, Amsterdam (1981).

24. I.B. Volkova. Characteristics of disperse organic matters in high-rank coals and associated rocks in Duniechi coalfield, Soviet Union (in Chinese), *Oversea Coal Geol.* 4, 7-12 (1990).

25. C. Wong and Z. Pan. Removal of the mineral matters in coals. *Bull. China Univ. Geosci.* 1, 214-221 (1982).

26. B.B. Xoxot. *Outbursts of Coal and Gas* (in Chinese), China Industrial Press, Beijing (1976).

27. Yang Qi. *The Advances in Coal Geology* (in Chinese). China Science Press, Beijing (1987).

28. Yang Shijing. Pore system and its characteristics of outbursted coals. In: *Proc. 2nd Int. Mining Congr.* Chinese Academy of Coal Science (Ed). pp.317-326, Beijing (1991).

Proc. 30th Int'l Geol. Congr., Vol. 18, Part B, pp. 123-133
Yang Qi (Ed.)
© VSP 1997

Advances of the Exploration and Research of Oil from Coal in China

HUANG DIFAN
Research Institute of Petroleum Exploration and Development, P. O. Box 910, 100083, Beijing, China

QIN KUANGZONG
University of Petroleum, P. O. Box 902, 100083, Beijing, China

Abstract

In recent years, great achievements have been made in exploring oil and gas from Jurassic coal measures in China. Plentiful oil and gas have been discovered in some coal basins of the Northwest China. The exploration, development and studies of the oil and gas from coal greatly improved the nonmarine oil generation theory. In this paper, a new method for the evaluation of oil and gas potential of coal and coal macerals is introduced, in which the organic carbon measured by ^{13}C NMR spectroscopy is distinguished into oil prone carbon (C_0), gas prone carbon (C_g) and aromatic carbon (C_a) with different chemical shift ranges. A primary migration model by three stages for oils from coal is also presented in this article.

Keywords: Oils from coal, ^{13}C NMR, Nonmarine oils, Primary migration

INTRODUCTION

The discovery of oil and gas field from coal and the study on its formation, expulsions and forming condition of pool in coal measures mark the most important advances in nonmarine oil generation theory.

Oil from coal is defined as liquid hydrocarbons formed by accumulated and/or dispersed organic matters in coal and coal measures during coalification. It may be expelled from hydrocarbon source rock of coal measures and accumulated to form pool, even large oil field, under special geological conditions and geological chromatographic effects. It is well known that coal measures may produce methane and result in large scale industrial accumulation of natural gas. However, traditional view held that environment for coal generation was not prone to oil generation. It was based on the fact that terrestrial oil source rocks are commonly lacustrine sediments, whereas coal measures are formed in swamp faces. Furthermore, the existing forms of organic matter are also different for lacustrine sediments and coal strata. Most oil fields do not coexist with coal fields in space.

From the late 1960's, some important oil and gas fields associated with Mesozoic and Cenozoic coal measures have been discovered in Gippsland Basin in Australia, Cuter Basin (Mahakam delta) in Indonesia, Scotia Basin and Mackenzie Basin in Canada, North

Sea Moray Basin in England etc.. Thus stimulate great interests in the study on oils fror
coal. In 1980's, extensive and thorough studies on oils from coal from many aspects hav
been made by the combination of organic petrology with organic geochemistry an
simulation experiments. A new field of oil from coal and its exploration in coal measure
is initiated following the theory of terrestrial oil generation. Now, it is widely accepte
that coal measures may form industrial oil pools, even large oil fields. A symposium on
Oil from Coal" held in Boston in April of 1990 sponsored by the Organic Geochemistr
Division of American Chemistry Society (papers were published in "Organi
Geochemistry" vol. 17, No. 6, 1991) reflects the extensive interest and advances in th
field of oil from coal.

EXPLORATION FOR OIL FROM COAL IN CHINA

In China, exploration for oil from coal has experienced a long time. In the past 30 years
some small oil fields were found in 50s to 60s, such as Yuanyanhu oil field in the wes
margin of Ordos Basin (Carboniferous-Permian), Zhongba condensate oil and gas field i
Chaidamu Basin (Jurassic), Qigu and Yiqikelike oil field (Jurassic) in pondment alon
north and south sides of Tianshan, Qiketai and Shengjinkou oil fields in Tulufan Basii
(Jurassic), etc. However, exploration in these coal measures has not brought significan
commercial benefit for a long time. The great breakthrough has been made in th
Shanshan arc structure belt of Tulufan Basin since 1989, discovering a series of oil field
associated with Jurassic coal measures. All these show a good prospect of large oil an
gas area. Other important discoveries have been made in the east of Jungger Basin, th
north of Tarim Basin, the east Basin of Jiuquan, the Santanghu Basin and Yanqi Basin o
central Tianshan. Especially in exploring the Tu-Ha Basin, the Jungger Basin and th
Kuche Basin. big oil and gas fields have been discovered related with coal measures, thu
showing good prospects for oil exploration in coal measures. Oils from coal measure
(J_{1+2} and P_1 - C_3) account for 3% of total oil and gas reserve in China and show a
increasing tendency.

Although both petroleum and coal are combustible organic deposits, coal differs from
petroleum fundamentally in their formation environments. The former is associated with
the accumulation of organic matter in swamp faces and the later is associated with th
concentration of disseminated organic matter in mudstone and carbonate of marine an
lacustrine faces. As a result, the ordinary coal producing area is not rich in oil and coa
beds poorly develop in oil-rich geological time. China is one of the countries abounding
in coal reserve, which is more than 900 billion tons. Coal is mainly derived from Middl
or Lower Jurassic and Upper Carboniferous to Lower Permian in China, which account
for 86.4% of the total coal reserve. On the other hand, according to the corresponding
distribution in age for different source rocks, oil and gas resources mostly concentrate i
Lower Cretaceous and Tertiary (mainly Lower Tertiary), accounting for 83.6% of the
total oil and gas reserve. It is beyond doubt that the accumulation of all these combustible
organic deposits occurs when paleoclimate is suitable and organisms are flourishing. It is

fferent geotectogenesis and sedimentary conditions that lead to the different existing
ates and mineralization conditions of the organic matter in sediments and the time
fferentiation between oil and coal reserves.

we have noticed, the formation conditions of oil and coal are not opposite completely
d are related to each other. Coal and oil both have organic origin. Coal appears in oil-
ch epoch and oil appears in coal-rich epoch. On the one hand, it is due to the coexistence
d alternation of these two sedimentary conditions; on the other hand, oil and gas will
evitably be formed in the process of coalification and asphaltization, i.e., coalification
cludes the process of hydrocarbon formation from organic matters. It seems that not all
nds of coal can form oil of commercial scale. The three essential conditions for the
rmation of an important commercial oil field from coal are: 1) Exinites in the coal
rmation swamps accumulated to some extent (more than 10% and may be lowered to
% at the stage of gas or fat coal, hydrogen index of kerogen $I_H > 100$ mg HC/ gC); 2)
sphaltization of the buried coal bed reaches gas coal to fat coal stage ($R_o < 1.5\%$); 3)
ne accumulation and reserve conditions of oil and gas are good. Oils from coal are
mmonly rich in gas and mainly appear in the form of light oils and condensates. It is
ne to the pore structural characteristics and high adsorptivity of the coal make the heavy
l difficult to be expelled. Mesozoic and Cenozoic coal-bearing basins are the major
aces for exploring oils from coal. Regarding the common fossil fuel energy resources in
e world, the ratio of coal to oil and gas is about 10:1, but the ratio is 50:1 in China. It
obably means that even more plentiful resources of oils from coal are waited for
xploration and development.

esults of our study on oil from coal are included in the books "Advances in
eochemistry of Oil from Coal" [1] and "Oil from Coal: Formation and Mechanism" [2].
'e summarized some achievement as follows.

VALUATION OF OIL AND GAS GENERATION POTENTIAL OF COAL BY C NUCLEAR MAGNETIC RESONANCE (NMR) SPECTROSCOPY

he evolution of chemical structure of coal mirrors the process of coalification. Solid
ate ^{13}C NMR spectroscopy plays an important role in studying chemical structure of
ganic macromolecules such as coal and coal macerals.

he chemical structure of various macerals and the richness of aliphatic carbon that is
rone to oil and gas generation in coal are different. According to the study of Qin et al.
,6], three carbon fractions of oil prone carbon (C_o), gas prone carbon (C_g) and aromatic
rbon (C_a) may be distinguished from the ^{13}C NMR spectrum of coal or coal maceral by
e help of deconvolution technique (Fig. 1). Based on the data of thermal simulation
xperiments, it is recognized that aromatic carbon in coal and kerogen contributes little to
ydrocarbon generation; whereas methylene and methine carbons are the main matrix of
l; other aliphatic carbons, carbonyl and carboxyl carbons are assigned as gas prone

carbon. A ternary diagram constructed from these three types of carbon (C_o, C_g and C) can be used to classify the type, to characterize the evolution and to access the oil and potential of coals, coal macerals and kerogens (Fig. 2).

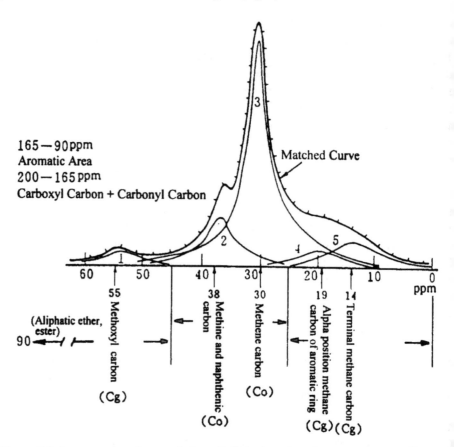

Figure 1 Polycomponent deconvolution of aliphatic carbon frequency band in ^{13}C NMR spectrum for kerogen from oil shale.

As shown in Figure 2, data of alginites and sapropellic coals concentrate in the upper part of the diagram, which corresponds to the region of type I kerogen. It indicates the alginite has a high potential of oil generation. Oil prone carbon fraction 0.65 may be taken as an average. Data of exinites including resinite, cutinite, sporinite and suberinite scatter in the middle part of the diagram, which corresponds to the region of type II kerogen. The hydrocarbon generation potential of resinite and cutinite is relatively high, but that of sporinite and suberinite is relative low, an average value of oil prone carbon of exinite may be taken as 0.35. Data of vitrinites and inertinites take place in the type III region, the lower left corner of the diagram. It seems reasonable to approximate their oil prone carbon fraction values as 0.04 and 0.02 respectively.

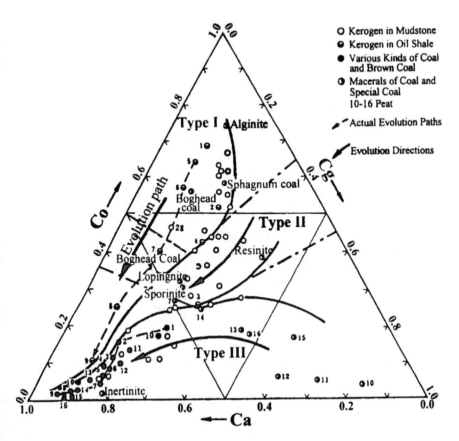

Figure 2 Types and evolution of three structural organic carbons of kerogen in source rocks, peat, various kinds of coal and macerals of coal.

If values of oil prone carbon fractions of coal macerals cited above are taken as weighted functions, the maceral composition of a coal sample may be used to evaluate its oil generation potential. For example, the brown coal from the Huangxian of East China is composed of 0.05 alginite, 0.16 exinite, 0.65 vitrinite and 0.14 inertinite [7]. According to the weighted functions, the contribution of the four macerals for oil generation is 0.033, 0.056, 0.026 and 0.003 respectively, having the sum 0.118. Hydrous pyrolysis thermal simulation experiment has been carried out for this brown coal under conditions of 350°C and 72 h. The ratio of carbon in yielded oil to the total organic carbon of the coal is 0.112 [6]. The experimental value agrees well with the above estimated value from the NMR diagram. The result indicates that exinites are the major contributor to oil generation. However, vitrinite should not be neglected because of its abundance. Analogous method may be used to evaluate the gas generation potential.

MODEL OF HYDROCARBON GENERATION AND EXPULSION OF COAL

The process of oil and gas migration is accompanied by hydrocarbon generation Zhong
analyzed 35 different coal samples of different coal rank in Tu-Ha Basin in order to stu
the hydrocarbon generation and expulsion process of coal. Based on the data of extrac
and group composition, the natural section for generation and expulsion of oil and g
from coal is illustrated in Figure 3.

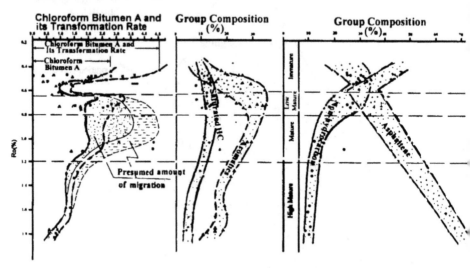

Figure 3 Natural hydrocarbon generation and expulsion section in Tu-Ha Basin.

As shown in Figure 3, the asphalt transformation rate presents two humps of hydrocarb
generating and evolutionary. The first hump occurred at R_o= 0.4-0.6% is assigned to th
early hydrocarbon generating and evolutionary peak of suberinite and vitrinite, accordi
to the identification of microscopic measurement. The products of this stage are immatu
oils. It is obvious that the second hump is the main hydrocarbon generating a
transforming peak because it occurs at R_o= 0.7-1.1%. However, the second peak
thought to be suppressed to certain extent according to the analysis of the peak shape, a
also the fact that the content of extractable bitumen and asphalt transform rate increa
1% and 1.5% only. This case is believed to be a phenomenon related to the loss during th
process of primary migration, which is also reported by Huc *et al.* in 1986 [3]. Based
the comparison of hydrocarbon generation and evolution of coal and mudstone, the
pointed out that the decreasing of hydrogen index in shales is matched by a correlati
increasing of the bitumen content (mg/gC) in sediments, however, this peak
extractable organic matter yield is not presented in coals, although the hydrogen index (I
of the coals examined are richer than that of the kerogens from shales. They conclud
that this phenomenon is resulted from the different expulsion ability, i.e., hydrocarbo
are more easily expelled from coals than from shales [3]. In this respect, coal, due to i
high organic matter content, will produce a higher absolute quantity of hydrocarbons
its pore system than organic matter dispersed in shales. The primary migration will b
favored for coals because the hydrocarbon saturation will reach more readily to th

minimum threshold of expulsion.

The evolution of group composition of extractable bitumen shown in Figure 3 also reflects obviously the regularity for the generation and expulsion of hydrocarbons. As $R_0 = 0.6\%$, it presents a low value of asphaltene content, a high peak of resins and an equivalent peak of aromatics. It indicates the occurrence of the primary expulsion related with the formation of immature oil.

With the evolution going on, non-hydrocarbons apparently transform into hydrocarbons and their content decrease dramatically to about 10%, while the asphaltene content increases significantly and may reach up to 70% at last. In the same time, the relative content of aromatics also decreases by about 15%, accompanied by a mild increasing of saturates by 2% or so.

From the group component evolution of the coal bitumen as stated above, some phenomena different from source rock of mudstone may be observed. They reflect the characteristics of migration and expulsion of oil and gas in coal:

(1) The expulsion of macromolecular asphaltenes from coal is difficult, leading to the linear increasing of its relative content with the maturation. This can demonstrate that most asphaltenes are retained in coal.

(2) With the increasing of maturation, aromatics decrease slowly, because only part of the mono-, bi- and tri-cyclic light aromatics may be expelled and most heavier aromatics, especially polycyclic hydrocarbons are still kept in the coal strata.

(3) During the whole process of evolution, the quantity of saturates is almost constantly lower than that of aromatics. The curve shape of the saturates changes mildly, showing a low content in the range of 5-15%, while the change of aromatics ranges between 15-30%. In the extract of source rocks, the predominance of aromatics over saturates is irregular in the respect of originally generated hydrocarbons; it cannot be explained by the humic nature of the matrix. This is because the most effective hydrocarbon generating components in almost any matrix are liptinites (or exinites), which generates primarily saturates but not aromatics. Therefore, this phenomenon can only be explained by the effect of migration. Since saturates are low in polarity, most of them have been expelled from the coal.

Because of the above characteristics of primary migration and expulsion of coal-type source rock , light oils are preferential for oils from coal gathered in reservoirs. The content of saturates in these oils may reach over 80%; mono-, bi- and tri-cyclic light aromatics are the next; heavy aromatics with more than four rings, non-hydrocarbons and asphaltenes rarely make actual contributions.

We have estimated the expulsion efficiency of coal by the normal saturates to aromatics ratio taken from the products from different types of kerogen. As to the saturates, their expulsion efficiency may reach 86-87% in the early and middle stage of hydrocarbon generation, and 91% in the later stage. The overall expulsion efficiency of saturates to the chloroform bitumen is 45-50%. This value is apparently lower than that in shales [4].

We suggested the primary migration model of oils from coal with three stages shown as Figure 4. The model is based on the distribution characteristics of the pores, micro fractures and the variation of internal water content of coal in the natural hydrocarbon generation and evolution section. It is also considered the effect of adsorptivity on the expulsion of generated hydrocarbons from coal and the phase variation of hydrocarbons in different stages.

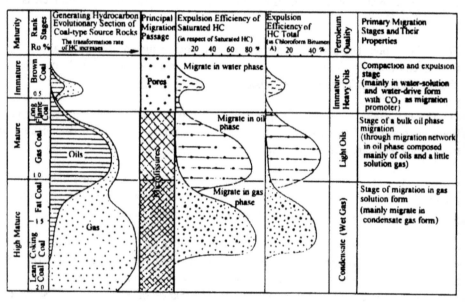

Figure 4 Primary migration model for oils from coal during coalification.

Compaction and expulsion stage

This stage occurs in the late diagenesis with vitrinite reflectance in the range of 0.4-0.7 %. In the corresponding coal strata, organic macerals such as resinite and suberinite contribute to the early generation of immature oils. In this stage, macropores (>30 nm) in coal are relatively common, making up more than 40% of the pore volume, which facilitate the expulsion of hydrocarbons. Moreover, this stage still belongs to the late diagenesis of intense compaction. The porosity of coal strata during this stage will decrease by about 10%. Accordingly, a great deal of internal water in coal pores (about 8%) will be expelled. Therefore, part of hydrocarbons and bitumen-like matters will be expelled or squeezed out in water solution and water-drive form. Under suitable conditions, they may form commercial accumulations of heavy immature oils. The characteristic of expulsion in this stage is the insignificant effects of geochromatograpy, since hydrocarbons are squeezed out with water. As a result, the accumulation of hydrocarbons still shows features of heavy crude oils with high content of nonhydrocarbons and asphaltenes. The relative density of oils is around 0.9 g/cm^3. Furthermore, the effect of carbon dioxide as a migration promoter for primary migration of oils should also be mentioned. Since coal is a humic organic matrix that can generate more carbon dioxide in the early hydrocarbon generation stage than common lacustrine mudstone source rocks.

Stage of bulk oil phase migration

It is the main hydrocarbon generation period (oil window) for organic matter in coal strata. All macerals of exinite play a role in hydrocarbon generation. The corresponding vitrinite reflectance is 0.7-1.2%, primarily equivalent to the gas coal stage. At this stage, the content of fluorescent vitrinite increases to 50-90%, demonstrating that vitrinite is widely disseminated by the mobile phase. The surface of mineral particles and non-crystalline organic carbonaceous matter is saturated by nonhydrocarbons and asphatenes, which facilitates the migration of saturates and part of light aromatics in a bulk oil phase. As to the driving force for migration, we think that the decrease and expulsion of the internal water are not important then, because the amount of the internal water may be expelled in this stage is less than 3%. During the process of hydrocarbon generation and transformation of organic matter, the volumetric expansion and abnormal high pressure (undercompaction) become the main driving force of migration.. It is significant because of transformation of organic matter from solid phase to gas and liquid phase. It is also caused by the temperature increasing (about 30 °C) due to the deepening of burial. As to the migration passage. the micropores in coal are developed in this stage while the macropores have been decreased to about 20% of the total pore volume and so play a minor role. In this period, the endogenetic fissures are developed considerably, about 4 cracks per centimeter in average. The exsudatinite is often produced along these cracks, which means that the endogenetic fissures make up the main passage of primary migration. Hence, in this stage, the combination of macropores and micro-fissures constructs the passage network for the migration of bulk phase, which is composed mainly of oils and a little solublized gas. Besides, coals of this stage are high in specific surface area and adsorption ability. As a result, quantities of non-hydrocarbons and asphaltenes lag behind in coal strata, while the generated hydrocarbons, which are possible to form commercial oil and gas accumulations, are the light oils with their density ranged from 0.80-0.84 g/cm^3.

Stage of migration in gas solution form

This is the main gas generation period when the evolution of hydrocarbon generation from organic matter in coal has reached the stage of high maturation. Heavy hydrocarbons and other organic matters of high carbon content will be cracked. The fluorescence of

vitrinites diminishes gradually, generating a great deal of gas and some light liqui
hydrocarbons. Thus, the gas solution form becomes the main way for the primar
migration of hydrocarbons. Therefore, hydrocarbons generated in this stage are wet gase
or condensate gases. As to the migration passage of this stage, following facts are thoug
to be important. The development of micropores is dominant with little macropores, whi
the development of endogenetic fissures and gas pores have reached the peak stage, up
8-9 fissures per centimeter. So microcracks nearly become the only primary migratio
passage, along which oils are expelled in the gas solution form. Expulsion efficiency
also the highest for saturates at this stage.

In summery, with the evolution of hydrocarbon generation during the process o
coalification, the phases and conditions for primary migration in coal strata develop an
change by stages: from the water phase migration, to the oil phase migration, and then t
the gas phase migration.

CONCLUSION

1. A series of commercial oil fields associated with Jurassic coal measures at northwes
China has been discovered in recent years, showing a good prospects for the exploratio
of oils from coal in China.

2. Based on the deconvolution of the ^{13}C NMR spectrum, oil prone carbon (C_o), ga
prone carbon (C_g) and aromatic carbon (C_a) of coals or kerogens can be distinguished.
ternary diagram composed by these carbon fractions is developed as a new method t
characterize and evaluate the oil and gas generation potential for coals and coal maceral
or kerogens.

3. According to the expulsion condition, the expulsion efficiency, the characteristic o
hydrocarbons from coal and the variation of phases, a new primary migration model b
stages of oils from coal is suggested.

REFERENCES

1. Huang Difan, Hua Axin, Wang Tieguang, Qin Kuangzong and Huang Xiaoming(Eds.). *Advances i
geochemistry of oils from coal,* Press of Petroleum Industry, Beijing (1992).
2. Huang Difan, Qin Kuangzong, Wang Tieguang and Zhao Xigu(Eds.). *Oil from coal: Formation an
mechanism,* Press of Petroleum Industry, Beijing (1995).
3. Huc A.Y., Durand B., Roucachet J., Vandenbrouck M. and Pittion J.C.. Comparison of three series o
organic matter of continental origin, *Org. Geochem.* **10**, 65-73 (1986).
4. Leythaeuser D., Mackenzie A. and Bjorøy M.. A novel approach for recognition and quantification o
hydrocarbon migration effects in shale-sandstone sequences, *AAPG Bull.*,**68**, 196-219 (1984).
5. Qin Kuangzong, Chen Deyu and Li Zhenguang. A new method to estimate the oil and gas potentials o
coals and kerogens by solid state ^{13}C NMR spectroscopy, *Org. Geochem.* **17**, 865-872 (1991).
6. Qin Kuangzong, Huang Difan, Li Liyun and Guo Shaohui. Oil and gas potential of macerals as viewed by
^{13}C NMR spectroscopy. In: *Organic geochemistry, poster sessions from the 16th International Meeting o
Organic Geochemistry.* Øygard K.(Ed) pp. 865-872. Falch Hurtigtrykk, Oslo (1993).

7. Qin Kuangzong, Yang qioushui, Guo Shaohui, Lu Qinghua and Shi Wei. Chemical structure and hydrocarbon formation of the Huangxian brown coal, China. *Org. Geochem.* 21, 333-341 (1994).
8. Zhong Ningning. Organic petrology study of oil from coal in Tu-Ha Basin. In: *Oil from Coal: Formation and Mechanism* T.G. Huang *et al.*(Eds) pp. 291-308. Petroleum Industry Press, Beijing (1995).

Proc. 30th Int'l Geol. Congr., Vol. 18, Part B, pp. 135-146
Yang Qi (Ed.)
© VSP 1997

Study on Jurassic Coal and Carbonaceous Mudstone as Oil Source Rocks in Tuha Basin, North-Western China

JIN KUILI, YAO SUPING, WEI HUI, TANG YAOGANG, FANG JIAHU and HAO DUOHU

Beijing Graduate School, China University of Mining and Technology, Beijing, 100083, P.R. China

ABSTRACT

The Early and Middle Jurassic terrestrial coal-bearing strata of Tuha basin, Xinjiang Autonomous Region, North-Western China (Fig. 1) are well-known coal and petroleum sources, especially a coal measure-related oil model from coal and carbonaceous mudstone in situ. This paper, based upon authors' research with comprehensive methods including sedimentology, organic petrology and geochemistry for a few years, shows ① that the coal measure-related oil generated in the main generating stage is earlier and originates from desmocollinite B, bituminite, cutinite and suberinite; ② that the oil-generating models of some macerals together with successful oil-expulsion experiment in terms of vitrain sample are expressed; ③ that the new laser-induced parameters and optical/morphological indicators of both CLSM and TEM for oil-source correlation can be proposed; and ④ that the sedimentary organic facies may be expressed in paleogeographic map, from which the quantity of coal measure-related oil, the difficult problem for geochemistry, may be solved.

Keywords: coal measure-related hydrocarbon, oil-generating model, oil-expulsion simulation, oil-source correlation, organic facies, paleogeographic map

INTRODUCTION

Now, the coal measure-related oil is still an investigation focus of fossil fuel geology. In China, since this kind of oil was discovered, it seems that the upsurge in its research has risen. Huang Difan et al. [4] in the light of geochemical research revealed coal-derived oil of this area and pointed out that "type III kerogen with a H/C ratio 0.7 - 1.0 can generate C_5-C_{21} liquid hydrocarbon as major products" and "the qualities of coal-derived oil are often better and lighter", but there is no answer to the problem where the coals and which macerals can generate oil.

In order to settle the "which" and "where" problems, we have done a large amount of research work for several years. First, based on the work of Huang Difan et al., simulation of hydrocarbon generation for individual macerals and organic facies samples, oil-expulsion experiments for vitrain and oil-source correlation (geochemical and photochemical methods) were done, in which some new techniques and a train of thoughts such as laser-induced fluorescence (photochemical technique), TEM and CLSM (Confocal Laser Scanning Microscope) used for oil-source correlation, pore volume and expulsion studies for oil migration etc. were developed.

To find out and confirm a certain amount of oil derived from Jurassic coal and carbonaceous mudstone, or precisely speaking, to evaluate petroleum quantitatively for sedimentary basin, the authors studied the ideas about organic facies from different authors, Rogers [10], Jones [8] and Huc [5]. We quite agree with Huc, because modern exploration requires a more quantitative approach. We modified Jones' organic facies and adapted them as mapping unit in paleogeographic reconstruction.

For the sake of saving space, a vast amount of sedimentological and geochemical data including oil-source correlation are not described in this paper.

GEOLOGICAL SETTING

Tuha and Junggar basins (Fig. 1) are postorogenic basins with an area of 50000 km^2 and 130000 km^2, respectively. Many Chinese researchers confirmed both of them to be a great flood basin as the status quo ante, i.e. during the Late Carboniferous to Early Permian, which was a sedimentary basin created due to collision of the surrounding plates. There had been sea invasion and volcanic pyroclastic event occurring in its early history, receiving continental deposit at the beginning of Upper Permian and broke up into two basins under subsequent collision in Late Mesozoic.

Strata of Tuha Basin are divided into two major sedimentary cycles [2], showing onlap $(P_2-J_2) \rightarrow$ offlap $(J_3-K_2) \rightarrow$ onlap (E-N) $\rightarrow$ offlap (Q). The main depression contains continental sediments larger than 5000 m. thick.

The lower and middle Jurassic coal-bearing strata vary in thickness from less than 1000 m. up to 2500 m. in the depocenter, contains mudstones, detrital rocks and more than forty coal seams, 100 m. or more in total thickness, mainly in Badaowan and Xishanyao Fms. and demonstrates lacustrine-swamp depositional system presenting three lower level sedimentary cycles.

From bottom ascending, the Jurassic stratigraphy is as follows:

Table 1. The Jurassic stratigraphy of Tuha Basin

Age	Formation	Thickness(m.)	Lithologic description	Main facies
J₃	Kelaza	355-655	Red-purple massive ss interbedded with sandy Mds	Fluvial
	Qigu	295-733	Chocolate sandy mds with oil-bearing ss	Fluvial
J₂	Qiketai	39-185	Yellowish green ss, sandy mds, black mds	Lake
	Sanjianfang	45-387	Brownish red and greyish green sandy mds	Lake
	Xishangyao	140-855	Upper, grey and brownish yellow mds, coal Middle, brown mds Lower, greyish white ss, coal	Swamp
J₁	Sangonghe	67-230	Brownish yellow mds	Lake
	Badaowan	43-850	Greyish white cong, mds, coal	Swamp

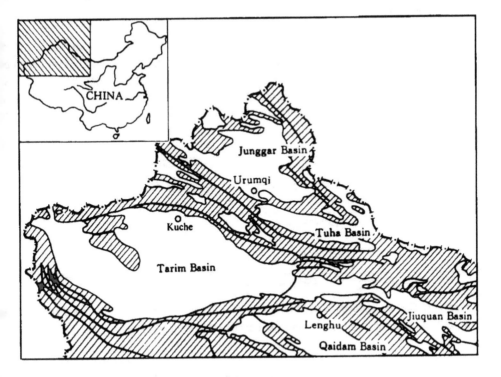

Figure 1. The location map of Tuha and Junggar basins.

As study objects, we not only paid attention to coal seams but also to carbonaceous mudstones, which ought to exceed 25% in maceral content in the whole rock estimate that is to say, they ought to be part of clastic marsh environment, and their organic richness might be 10% roughly (based on TOC=DOM × 0.5 × (0.8-0.85)) [11].

OIL-GENERATING MACERALS AND MAIN STAGE OF OIL-GENERATION

The maceral composition of Tuha coal is that exinite, inertinite and vitrinite may b 6-8%, 10-20% and 70-80%, respectively, of which desmocollinite proportion is ver; high, generally making up 25-45% in the total coal. The maceral composition of so called carbonaceous mudstone in this paper is similar to coal but disseminated and fractured. Based on fluorescence, TEM and micro-FTIR investigations, th desmocollinite can be divided into two types, Type A and Type B. The latter is th dominant and has fluorescence, rich in submicroscopic exinites and good potential o oil generation, with S_1+S_2 up to 200-300 mg/g, and may change into bituminite, al these characteristics are contrary to the former.

The main oil-generating macerals of coal and carbonaceous mudstone (see Table 2, 3, for liquid hydrocarbon are mainly desmocollinite B, bituminite, cutinite and suberinite which would have generated liquid hydrocarbon in low rank (VRr = 0.4 - 0.6% according to our study on hydrocarbon generated indications (e. g. oil drops, oil film and micrinites). The coalification map of the lower part of coal measures is as follow: (Fig. 2).

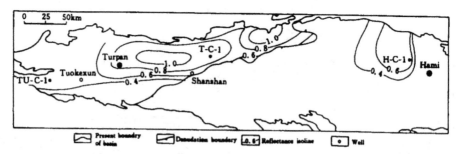

Figure 2. The top surface's coalification map of Badaowan Formation (J_1^1), modified from the Research Institute of Petroleum Exploration Development, Beijing (1996)

We examined oil samples from this basin with CLSM and TEM, and discovered some vitrodetrinites under CLSM. The reflectance of these vitrodetrinites is similar to that of coal or carbonaceous mudstone in coal measures of this basin; in addition, a lot of submicromacerals and Jurassic microfossils were discovered under TEM (Fig.3 and Fig.4). The conventional oil was short of these evidences.

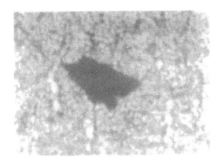 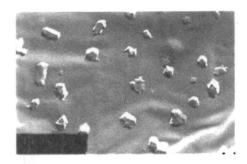

Figure 3. Vitrinite relic in oil from Tuha basin,under CLSM, × 3300, strew slide

Figure 4. Vitrinite relics in oil from Tuha basin, under TEM, × 33000, replica of oil

THE LASER-INDUCED FLUORESCENCE METHOD USED IN OIL-SOURCE CORRELATION

In order to verify the above conclusion, different oils and aromatic fractions from source rocks including coal and carbonaceous mudstone were used for correlation by indicators, such as maturity indicator, standard compound, fluorescence spectrum and fluorescence lifetime fingerprint, from which the so-called laser-induced fluorescence method was proposed by us [6,7] in addition to the use of conventional geochemical method for correlation.

From both results (Fig. 5), it is not only proven to be true for coal measure-related oil, but also manifested itself in some aspects of laser-induced fluorescence pattern, at least, the decreasing trend in both peaks and lifetimes, may be characteristics of coal measure-related oil, i.e., the condensate and light oils.

HYDROCARBON GENERATION MODELS FOR INDIVIDUAL MACERALS

We have made thermal simulation experiments using high Pressure Vessel under 6 temperature (150℃-330℃) conditions (Table 2 and Fig. 6) and quartz tube under 7 temperature (200℃-400℃) conditions and analyzed by using PY-GC for the former, the micro-FT-IR as well as fluorescence microscopy for the latter. The results not only confirm that the above mentioned macerals can generate oil at the early stage, but also show individual maceral's oil-generating models (Fig. 7). In the meantime we use factor A and so on [1] for study.

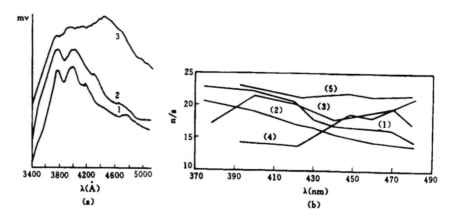

(a)

(b)

Figure 5. Oil-source correlation used in Tuha basin

(a) Oil-source rock spectra: (1)carbonaceous mudstone; (2)coal-generating oil; (3)conventional oil.
(b) Oil-oil lifetime fingerprints: (1)coal-generating oil; (3)Tuha's conventional oil; (4), (5)Tarim
conventional oils

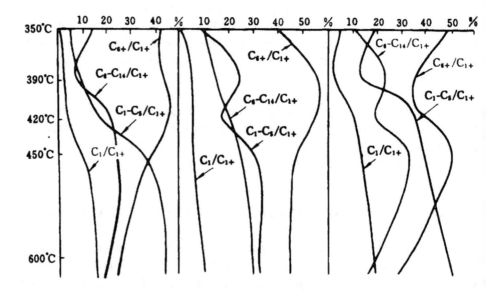

Figure 6. The evolution diagram of hydrocarbon parameter by PY-GC study

Table 2. Oil yields from simplified simulation results of pressure vessel

Maceral	Gas-generating yield (ml/g)	Pyrolysis oil yield (mg/g)	Heavy oil yield (mg/g)	Total liquid hydrocarbon yield (mg/g)
Desmocollinite	124.93	16.4	15.81	33.83
Cutinite	423.95	164.6	35.96	200.6

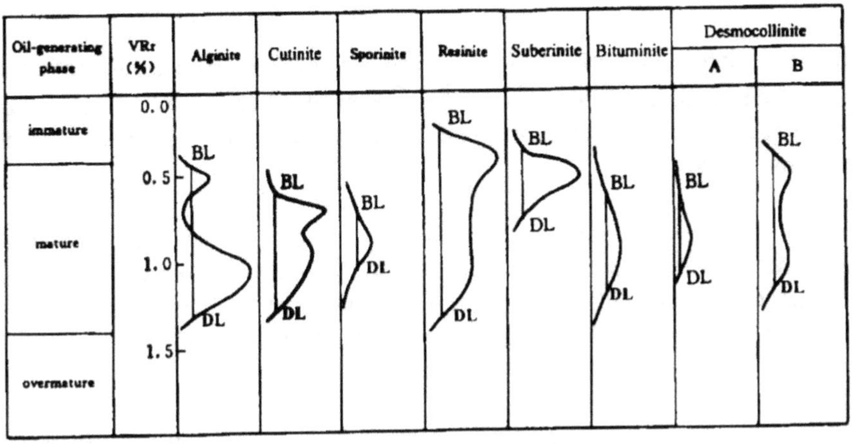

Figure 7. The oil-generating models of individual macerals in Jurassic coal measures

Alginite(Pila): During the thermal evolution, the obviously variable characteristics of alginite, telalginite from pila is peak variation at 2950 cm-1 and 2960cm-1, up to the maximum strength at the temperature of 290℃, disappearing at the temperature of 400 ℃. Factor A (2860cm-1+2930cm-1) / (2860cm-1+2930cm-1+1600cm-1) gradually increases below the temperature of 290℃, and decreases above the temperature of 290 ℃, Kal (2930cm-1+2860cm-1) increases at the beginning and decreases in the final. The variable characteristics of factor A and Kal indicate that alginite enters the "oil-generation window" relatively later.

Desmocollinite: Kal and factor A of desmocollinite A have only a limited value, with the reflectance being 0.9~1.1% VR$_0$ or so. However, Kal and factor A of desmocollinite B have two extreme values, 0.5% VR$_0$ and 0.9% VR$_0$ respectively, which indicates desmocollinite B generates hydrocarbon earlier and terminates later, with two hydrocarbon-generating peaks. And at the same maturity, aliphatic structure absorbency intensity of desmocollinite A is much less than that of desmocollinite B, indicating the hydrocarbon-generating potential of desmocollinite A is less than that of desmocollinite B.

Bituminite: It contains abundant CH_2 and CH_3 functional groups. With the increase of the thermal evolution, Kal of bituminite reaches the maximum value at 260℃~290℃ and becomes weak at 400℃. The peak at 1460cm-1 has the maximum value at 290 ℃.The above result indicates that bituminite generates hydrocarbon relatively earlier.

Sporinite: The maximum value of Kal and factor A is in the range of 290℃~320℃, showing that in this temperature range the liquid hydrocarbon yield of sporinite is up to the maximum.

Cutinite: At 260℃, the peaks at 2950cm-1 and 2860cm-1 representing aliphatic radicals are up to the maximum value, and still keep relatively strong at 360℃, indicating that cutinite starts generating oil earlier and terminates later, with a wide range of oil-generation window.

Resinite: In the range of 0.4~0.6% VR_o, the intensity of stretching vibration absorbency of CH_2 and CH_3 at 2950cm-1 and 2860cm-1 obviously decreases, and continues to decrease with the increase of temperature. The above result indicates that resinite is characterized by early hydrocarbon generation and wide range of oil-generation window up to 1.1% VR_o or more.

Modern subereous tissues (substituted for suberinite): The peaks at 2950cm-1 and 2860cm-1 reach the maximum values at 200℃, soon afterwards these band strength decreases. The aliphatic structure absorbency disappears at 320℃. The above result indicates that modern subereous tissues generate hydrocarbon earlier and terminate earlier. It starts at about 0.5% VR_o to generate a large amount of hydrocarbon, and terminates at 0.9% VR_o or so.

OIL-EXPULSION EXPERIMENT

Now, geologists focus their attention upon problem of oil-expulsion from coal. Coal macropore volume of extracted and unextracted vitrain samples were compared based on mercury pressure porosimetry together with the TEM and SEM. The results display that the total pore volume increases after extraction and the pore connection may be in series/parallel pattern except the isolated pores. That is to say, oil can be expelled. Moreover, the oil-expulsion experiment which was done in 72 hr's., below 210℃ and 18 atm. was further studied. It confirmed the above conclusion with evidence that the expelled oil is what is pressed into the coal (Fig. 8). The experiment is as follows:
vitrain sample → drawing air → soaking the vitrain sample in oil → drawing air →

ressed in pressure vessel → expelling oil from vitrain sample → PY-GC.

SEDIMENTARY ORGANIC FACIES

Organic facies is first proposed by Rogers [10], and Jones [8] defined it as a mappable subdivision of a designated stratigraphic unit. Based upon this idea and especially that of Huc's [5], a key organic facies to improve quantitative petroleum evaluation, we emphasize sedimentary parameters including maceral and palynofacies analyses in addition to lithofacies study for the sake of putting them into paleogeographic map, and call the organic facies proposed by Rogers and Jones as sedimentary organic facies. Therefore, we have revised the coal facies that contains coal/mudstone (>25% maceral content) as swamp/marsh facies and further subdivide it in view of sedimentary system, of which the inland plain have lake-swamp (marsh) and river-swamp (marsh) systems. Four sedimentary organic facies of coal and carbonaceous mudstone were suggested: namely high moor, forest swamp/marsh, running water swamp/marsh and open water facies (Table 3). The term of running water facies originates from C.H.Haymoba [3].

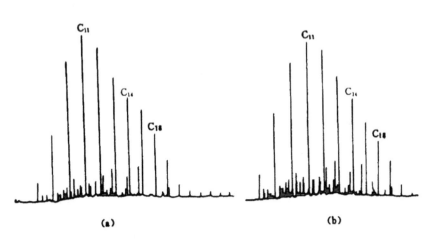

Figure 8. GC spectrum of pressed-in oil (a) compared with that of the expelled (b)

Because macerals contain dual natures of both petrologic and geochemical characters, we may quickly set up organic facies based on maceral statistics. The organic geochemical parameters of some macerals are listed in Table 4, the sum of S_1+S_2 converted based upon hydrocarbon potential of oil-generating macerals is closer to the actual measurements of Rock-Eval. Similar conclusion in dividing sedimentary organic faces has been drawn using the above two methods. The running water swamp facies zone may be the best for coal-generating oil (Fig. 9).

Table 3. The Classification of the Sedimentary Organic Facies

Facies Marks	Type Number	High moor facies	Forest swamp / marsh facies	Running water/ marsh facies	Open water facies
		1	2	3	4
Organic petrologic character	V+I%	>90	70-90	40-70	<40
	E%	0-10	10-30	30-60	>60
	Main maceral	Fusinite	Telocollinite	Desmocollinite	Alginite, Cutinite
Sedimentary character	Microlithotype	Fusite	Vitrite, clarite	Clarite	Durite
	V/I	<1	>1	>1	>1
	GI	1~2	2~50	0~50	2~10
	TPI	0~2	2~6	0~2	2~10
Organic geochemical character	H/C	<0.95	0.95-1.15	1.15-1.4	>1.4
	HI(mg/g COT)	<125	125-250	250-400	>400
	S_1+S_2(mg/g)	<50	50-200	200-300	>300
	Organic matter type	III	II B	II A	I
Jones' organic facies		D, CD	C, BC	BC, B	B, AB, A

Table 4. Macerals' hydrocarbon Potential (S_1+S_2)

Maceral	VRr(%)	S_1 (mg/g)	S_2 (mg/g)	S_1+S_2 (mg/g)
Cutinite	0.5	9.91	358.23	368.14
Desmocollinite B	0.48	6.08	245.74	251.82
Suberinite	0.44	3.90	140.8	144.7
Mineral-bituminous groundmass	0.51	0.63	9.03	9.66
Bituminite	0.51	9.1	439	448.1

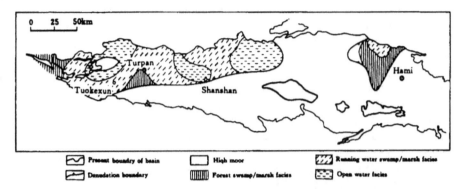

Figure 9. The sketch map of sedimentary organic facies of the Badaowan stage (J_2^1) in Tuha Basin

From this map, it is easy to settle the problems "which" and "where" the coal measure-related oil comes from; moreover, the quantity of oil resource can be obtained.

CONCLUSIONS

Based upon our three-years work, the main aspects concerning oil-generating coal measures of Tuha basin are exposed. Not only exinites but also desmocollinite may play an important role in oil generation, earlier than birth line of conventional oil or so. Confirmation of coal measure-related oil and success of oil-source rock correlation by means of CLSM, TEM and laser-induced fluorescence spectrometry may act as new techniques more cost-effective and quicker than geochemical method. Audio-visual oil-expulsion experiment may serve as evidence for Levine's theory [9] which points out pore system being without "plugging" with hydrocarbon at low rank coal. Individual macerals' model for oil generation and sedimentary organic facies result improve the theoretical answers to "which" and "where" the coal measure-related oil comes from.

From oil-generating coal measures' data worldwide, it seems that the running swamp/marsh is developed on limnic environment frequently, therefore, there is an allochthonous/hypautochthonous process as well as disintegration, which causes accumulation of macerals rich in hydrogen content. When coal and carbonaceous mudstone have enough high hydrogen contents relative to carbon, they can form oil earlier than the birth line of conventional oil generation or so.

REFERENCES

1. Ganz H.H., Kalkreuth W. Application of infrared spectroscopy to the classification of kerogen types and the evaluation of source rock and oil shale potentials. *Fuel.* 66,708-711(1987).
2. Geological Team 1, Xinjiang Bureau of Geology and Mineral Resources. Origin and development of Turpan-Hami Coal Basin and Accumulation for coal seams. Science-Medicine Press, Urumqi. 1(1992). (in Chinese).
3. Haymoba C.H. The genesis classification for coal of Suburbs-Basin of Moscow. *U.S.S.R. Mineral Resources Institute Bull.* 159 (1940). (in Russian).
4. Huang Difan, Zhang Dajiang, Li Jinchao and Huang Xiaoming. Hydrocarbon genesis of Jurassic coal measures in the Turpan Basin, China. *Org. Geochem.* 17:6, 827-837 (1991).
5. Huc A.Y. Understanding organic facies: A key to improved quantitative petroleum evaluation of sedimentary basins. In: Organic Facies. A.Y. Huc (Ed.). A.A.P.G. Tulsa, Oklahoma, 1-11(1990).
6. Jin Kuili and Qiu Nansheng. Application of laser-induced fluorescence of coal extracts for determining rank. *Org. Geochem.* 20:6, 687-694(1993).
7. Jin Kuili, Fang Jiahu, Guo Yingting, Zhao Changyi and Qiu Nansheng. The use of laser-induced fluorescence indicators in determining thermal maturation of vitrinite-free source rock and oil-source rock correlation (abs.). *Twelfth meeting of the Soc. for Org. Petrol.* 12, 18-22(1995).
8. Jones R.W. Organic facies. In: *Advance in Petroleum Geochemistry.* J. Brooks and D. Welte (Eds). Springer-Verlag, Berlin. 2, 1-90 (1987).
9. Levine J.R. Relationship to the molecular fraction of coal to measurements of porosity and density. Implication regarding the role of coal as a petroleum source rock and reservoir (abs.). *Ninth annual meeting of the Soc. for Org. Petrol.* 2(1992).
10. Rogers M.A. Application of organic facies concepts to hydrocarbon source rock evaluation. *Proc. 10th World Petr. Cong.* 2, 23-30 (1980).
11. Smyth M., Cook A.C. and Philp R.P. Birkhead Revisited: Petrological and geochemical studies of the Birkhead Formation, Eromanga Basin. *APEA Jour.* 2:1, 230-242(1984).

Proc. 30th Int'l Geol. Congr., Vol. 18, Part B, pp. 147-159
Yang Qi (Ed.)
© VSP 1997

Reaction Kinetics of Coalification in the Ordos Basin, China

LIU DAMENG, YANG QI and TANG DAZHEN

Department of Energy Resources and Geology, China University of Geosciences, Xue Yuan Road

29, Beijing 100083, P.R.CHINA

Abstract

Research on the reaction kinetics of coal rank suites of various ages in the Ordos basin has shown that the average apparent activation energy of coal is relatively high, ranging 0.42%-0.53% $R_{o,m}$, and the distribution of activation energy is rather dispersive; the average apparent activation energy of rank 0.58%-1.73% $R_{o,m}$ and the activation energy at the stage of the highest hydrocarbon generation increase with the increasing coal rank, and the activation energy distribution turns from dispersion to convergence; the Triassic coal has the highest activation energy and the strongest tendency in hydrocarbon generation compared with the corresponding coal ranks of other ages; the essential structure of coal is characterized by its high degree of aromatization at the high metamorphic bituminous stage($R_{o,m} > 1.73\%$), the apparent activation energy of coal displays as low values, and the distribution of activation energy is dispersive again due to the insufficient pyrolysis of the instrument. In addition, the kinetics analysis also indicates that the oil-gas accumulation from coal-derived hydrocarbons are mainly pyrolyzed hydrocarbons, while cracked and biogenetic hydrocarbons are relatively rare; identical rank coals can be formed under long-term and low-temperature or short-term and high-temperature; the heating rate and the degree of thermal evolution restrict the T_{max} values altogether, and the pyrolyzed products affect the T_{max} values as well.

Keywords: Ordos basin, coalification, activation energy, coal pyrolysis, reaction kinetics

INTRODUCTION

Coalification plays for a long time an important role in evaluating the coal quality, directing the rational. utilization of coal, as well as predicting and exploring the necessary coal types, etc. [20-22]. In recent years, the rapid development of organic petrology has deepened the study of dispersed organic matter in sedimentary rocks, which greatly widens the study scope of coalification, and promotes its significance. The study is concerned not only with coal quality, but also helps to solve other relevant geological problems in stratigraphy and structures, reconstructing the paleogeotemperature history of the basin, the paleogeography and the paleotectonics, especially the contrasting of the coalification stage with the oil-gas maturity which makes coalification to be extensively applied to long-range predicting and exploration of oil-gas-bearing areas.

The study on coal structure shows that coal may be regarded chemically as an inhomogeneous polymer with three-dimensional network. A series of complicated chemical reactions on the chemical composition and structure of coal occur during coalification. Learning all the above, people try to reveal the coalification mechanism through studying the structural characteristics of coal. It is found that the coalification process can be described by using a series of parallel first-order reaction[1,2,6,8,11-13]. Many types of chemical bonds known in coal are broken off in order of low to high bond energy with increasing temperature. As to hydrocarbon generating from the breakage of each type of bond, the first-order reaction under constant temperature may be used to cope with it, while the gross process can be regarded as the algebraic sum of a series of parallel first-order reactions. According to the reaction activation energy of a set of bond types, the concentrations of a set of bond types in coal or kerogen can be calculated, and the hydrocarbon generation potential of coal or kerogen can be calculated as well. Many Chinese scholars calculated the oil-generating amount by using chemical reaction kinetics, and put forward some profitable methods and formulae[9,14,15,17-19,23,24]; Wang[16], Wu[19], Jin[5] and Luo[10], etc. calculated the oil-gas generating amount of some basins, hydrocarbon generation rate in different buried depth, temperature and depth of oil generation threshold by using the combination of chemical reaction kinetics model with the geological parameters such as depositional rate and geotemperature gradient. This paper tries to recognize the kinetics character of various rank coals, to probe into the coalification mechanism and to reveal the geochemical significance for the average apparent activation energy and activation energy distribution of coal pyrolysis.

THEORETICAL BASES OF REACTION KINETICS OF COAL PYROLYSIS

The gaseous, liquid and solid (remnant carbon) products are always formed when coal pyrolyzed. The liquid product is quickly volatilized and separated from reaction area. So, the rate that formed the volatile matter may approximately be considered to be the coal pyrolysis rate. For these reasons, the coal pyrolysis model can be simplified as follows:

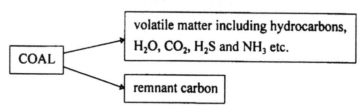

According to the basic principles of chemical reaction kinetics, the relationship between the reaction time, the concentration of reaction matter and formation rate of reaction product can be described as follows:

$$dx/dt = k(1-x)^n \tag{1}$$

From the Arrhenius formula, the reaction velocity constant K in formula (1) is exponentially related to activation energy:

$$k = A\exp(-E/RT) \tag{2}$$

Substitute formula (2) into formula (1):

$$dx/dt = A\exp(-E/RT)(1-x)^n \tag{3}$$

In formula (3), t represents time, second; x, volatile concentration formed at the reaction time t; n, reaction order; K, reaction velocity constant(S^{-1}); A, pre-exponential factor(S^{-1}); E apparent activation energy of reaction matter (kJ/mol); R, gas constant (kJ/°C mol); T, reaction temperature(°C).

In experiment constant heating is adopted, that is, $dT/dt=C$; substitute it into formula (3):

$$dx/dT = A/C\exp(-E/RT)(1-x)^n \tag{4}$$

As stated above, the first-order reaction is applicable in a relatively large scope of x, consequently the formula (4) can be simplified as follows:

$$dx/dT = A/C\exp(-E/RT)(1-x) \tag{5}$$

The formula (5) is the basic kinetics equation of coal pyrolysis.

EXPERIMENTAL SAMPLES AND METHODS

All coal samples were collected from four sets of coal-bearing strata in the Ordos basin, which are respectively the Taiyuan Formation(C_{3t}), the Shanxi Formation(P_{1s}), the Yanchang Formation(T_{3w}) and the Yan'an Formation (J_{1-2y}), their vitrinite reflectance being 0.42%-2.11% $R_{o,m}$. In order to measure relatively accurately the activation energy of coal, and to exclude the disturbance from coal samples, the hand-picked banded vitrains in bright coal are adopted as experimental samples(Table 1).

Table 1. Geological characteristics of coal samples for pyrolysis kinetics analysis

Sample No.	Sampling Site	Age	$R_{o,m}$/%	Sample Characteristics
1	Dongsheng,Inner Mongolia	J_{1-2y}	0.42	banded vitrain
2	Chengpo,Inner Mongolia	P_{1s}	0.53	banded vitrain
3	Hequ,Shanxi	P_{1s}	0 58	banded vitrain
4	Pianguan,Shaanxi	C_{3t}	0 59	banded vitrain
5	Yuling, Shaanxi	J_{1-2y}	0.65	banded vitrain
6	Xingxian,Shanxi	C_{3t}	0.71	banded vitrain
7	Hongshiya,Shaanxi	J_{1-2y}	0.75	banded vitrain
8	Zichang,Shaanxi	T_{3w}	0.80	banded vitrain
9	Pangpangta,Shanxi	C_{3t}	0.95	banded vitrain
10	Yuming,Shanxi	P_{1s}	1.02	banded vitrain
11	Liuling,Shanxi	P_{1s}	1.42	banded vitrain
12	No.2 Jingao mine,Shanxi	P_{1s}	1.68	banded vitrain
13	Maozequ,Hejing	P_{1s}	1.73	banded vitrain
14	Xiangshan mine, Hancheng	P_{1s}	1.99	banded vitrain
15	Xiangshan mine, Hancheng	C_{3t}	2.11	banded vitrain

Rock-Eval II pyrolytic instrument was used in the experiment with the start temperature at 200°C, the constant heating pyrolysis was done at the heating rate of 10 °C /min, 20 °C /min, 30 °C /min, 40 °C /min and 50 °C /min, and the highest temperature at 600 °C. The instantaneous cumulative area of the peaks of pyrolyzed hydrocarbon (S_2) °C was typed by using integrator (at the heating rate of 10 °C /min, typed once every 10 °C; at the other heating rates, typed once very 20 °C). Each instantaneous hydrocarbon generation rate (at each temperature point) was calculated using each instantaneous typing value and total area of peak S_2.

Because the temperature and reaction rate dx/dt of the same sample reaching the same hydrocarbon generation rate under varying heating are different, substitute T and dx/dt at the same rate of hydrocarbon generation and varying heating rate into formula (5), A and E can be calculated by using the least square equation.

EXPERIMENTAL RESULTS AND DISCUSSIONS

Kinetics Parameters Vary with Coal Rank

Judging from Figure 1 and Table 2, at the range of 0.42%-0.53% $R_{o,m}$, the average apparent activation energy is relatively high; when the coal rank increases to 0.95% $R_{o,m}$, the average apparent activation energy displays low value; in the range of 0.95%-1.42% $R_{o,m}$, the average apparent activation energy increases with the increasing coal rank; afterwards, the average apparent activation energy decreases with increasing coal rank. The variation tendency of the pre-exponential factor A with coal rank is similar to that of the average apparent activation energy. It should be pointed out that the coal of Late Triassic Yanchang Formation shows the highest activation energy and the strongest hydrocarbon generation trend compared with the corresponding rank coals of other ages in the basin(Table 3), which is possible related to the fact that Triassic coal is relatively rich with lipids in microscopic chemical component. According to studies by Zhang[24] and Wang[15], the apparent activation energy of various types of kerogens shows $E_I > E_{II} > E_{III}$ during the whole oil generation stage, which indicates that the higher the content of aliphatic structural component is, the higher the activation energy. This is because the pure humic kerogen (including coal) is characterized by a uniform

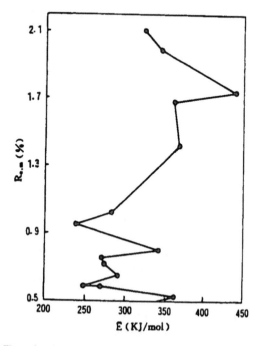

Figure 1. Average apparent activation energy varies with coal rank

weight loss, and the side chain of condensed aromatic nuclei cracking gradually, while the pure sapropelic kerogen (Type I kerogen) is characterized by a long chain of lipid structural unit being bridged with the heteroatomic radicals of oxygen and sulfur, etc. The activation energy between large radicals is relatively high, characterized by an instantaneous weight loss, and large amount of oil-gas which is easily generated under the appropriate conditions, such as proper temperature and time etc.[3,4].

Table 2. Pyrolysis kinetics parameters in various rank coals

$R_{o,m}$(%)	$A(S^{-1})$	$\bar{E}$ (kJ/mol)	R
0.42	2.44×10^{30}	250	0.986
0.53	3.41×10^{31}	365	0.877
0.58	2.3×10^{18}	270	0.986
0.59	5.58×10^{19}	250	0.969
0.65	4.78×10^{19}	291	0.997
0.71	2.52×10^{23}	275	0.976
0.75	8.04×10^{22}	271	0.998
0.80	5.57×10^{28}	341	0.988
0.95	1.91×10^{18}	239	0.985
1.02	3.3×10^{20}	282	0,.987
1.42	5.04×10^{31}	368	0.988
1.68	4.02×10^{33}	361	0.991
1.73	6.74×10^{36}	439	0.984
1.99	4.3×10^{35}	344	0.938
2.11	7.38×10^{33}	323	0.994

Table 3. Relationship among hydrocarbon generation window, activation energy distribution intervals and rate of maximum hydrocarbon generation with heating rate 20 °C /min

$R_{o,m}$(%)	T_{10}(°C)	T_{90}(°C)	ΔT(°C)	ΔE(kJ/mol)	X_{max}(%)
0.42	366.7	516.7	150	383.12	21.77
0.53	369.7	504.2	134.5	209.82	56.73
0.58	369.9	480.5	110.6	138.52	74.11
0.59	362.5	468.1	105.6	123.17	68.31
0.65	369.7	463.9	94.2	113.99	79.31
0.71	366.7	475	108.3	163.39	51.47
0.75	376.4	480.5	104.1	164.27	59.94
0.80	375	481.9	106.9	212.94	75.67
0.95	366.7	468.1	101.4	111.32	47.87
1.02	367.6	481.5	113.9	174.07	80.58
1.42	380.5	505.6	125.1	251.60	75.81
1.68	398.2	526.4	108.2	387.53	75.67
1.73	416.7	527.8	111.1	287.04	69.21
1.99	422.2	540.3	118.1	439.58	29.45
2.11	418.1	543.1	125	431.46	45.36

T_{10} indicates the corresponding temperature at the hydrocarbon generation rate of 10%; T_{90}, the corresponding temperature at the hydrocarbon generation rate of 90%; ΔT, the difference of temperature between T_{90} and T_{10}; ΔE, the activation energy intervals at the corresponding temperatures; and X_{max}, the maximum rate of hydrocarbon generation.

The above-mentioned reasons for variations of kinetics parameters can be interpreted as follows: at the immature stage, when $R_{o,m} < 0.53\%$, immature hydrocarbons are formed due to a large amount of nonhydrocarbon gases such as CO_2, H_2O, H_2S, N_2 and CO which are produced owing to the presence of large amount of heteroatomic radicals and biomarkers in coal during pyrolyzing, thus resulting in maturation and metamorphism of coal at the same time; therefore, the average apparent activation energy exhibits relatively a high value. In the range of $0.58\%\text{-}1.73\%R_{o,m}$, the cracking of side chain, the coming off of functional groups and aromatization of structural unit are predominant,

C — H bonds $\quad$ H₃C—H $\quad$ 423–436
$\qquad\qquad\qquad$ C₂H₅—H $\quad$ 406–410

C — C bonds

544 | 350 $\quad$ ⬡—CH₂—⬡
480 | 349 $\quad$ H₃C—CH₃
382 | 235 $\quad$ ⬡—CH₂—CH₂—⬡
364 | 210 $\quad$ ⬡⬡—CH₂—CH₂—⬡⬡

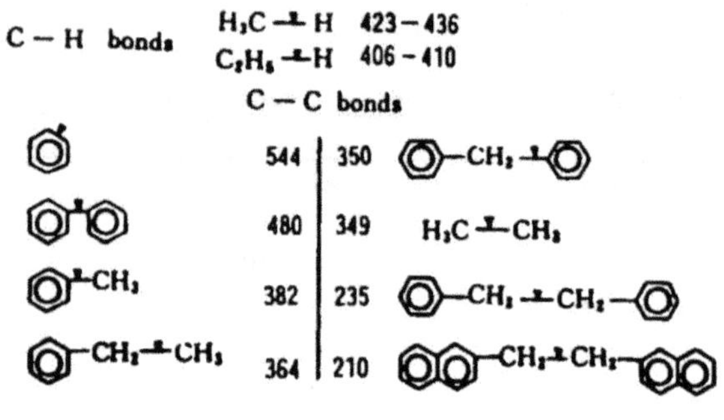

Figure 2. Bond-energy values of hydrocarbons (kJ/mol)[7]

thus bringing about the increase of aromatics and decrease of aliphatics, while the cracking of C-C bond between aromatic rings is more difficult than that of C-C band between aromatic ring systems or aliphatics(Fig.2). Hence, the average apparent activation energy and activation energy at the stage of maximum hydrocarbon generation increase with coal rank. At the high metamorphic bituminous to anthracite stage, condensation predominates, the essential structure of coal is characterized by a high degree of aromatization, but the highest pyrolyzing temperature of the instrument reaches only 600 °C, which is insufficient to cause a complete pyrolysis of coal, and the average apparent activation energy displays as low values again.

Table 4. Relationship between hydrocarbon generation rate and apparent activation energy in various rank coals

$R_{o,m}(\%)$ \ $Xi(\%)$	5	10	20	30	40	50	60	70	80	90
0.42	128.92	136.51	162.79	164.75	174.36	199.33	206.52	246.46	268.66	519.63
0.53	296.75	311.29	275.64	275.09	286.61	279.51	306.28	369.33	409.81	521.11
0.58	211.37	212.69	210.76	212.31	211.96	210.09	228.29	246.29	273.19	351.21
0.59	214.25	218.63	207.89	196.43	201.00	200.04	207.59	230.67	257.25	341.80
0.65	221.18	225.82	224.49	221.79	215.79	216.72	224.39	241.64	277.33	339.81
0.71	197.90	235.14	237.22	230.30	228.89	235.12	245.14	263.62	303.03	398.53
0.75	226.44	226.15	230.61	229.37	220.88	227.26	231.82	254.15	287.62	390.42
0.80	284.47	262.98	234.17	223.36	211.69	217.59	231.11	263.57	321.81	475.92
0.95	177.80	207.94	190.44	209.55	212.05	203.64	211.21	209.56	227.94	319.26
1.02	206.48	182.30	195.65	233.66	255.20	236.30	224.28	230.89	261.47	356.37
1.42	293.65	280.84	260.05	234.24	237.07	234.43	237.12	251.70	307.05	532.44
1.68	317.97	189.86	203.68	227.01	213.82	227.23	232.33	257.61	331.70	577.39
1.73	382.93	331.58	270.61	295.06	300.55	305.41	320.98	362.64	494.85	618.62
1.99	249.71	236.27	253.36	257.68	264.09	279.09	297.60	343.31	434.44	675.85
2.11	184.41	230.95	231.54	235.34	237.34	238.41	266.70	326.83	444.83	662.41

Table 4 shows that the apparent activation energy increases with the hydrocarbon generation rate
X, and it increases linearly and varies slowly at the hydrocarbon generation rate of 10%-80%,
while the hydrocarbon generation rate is more than 80% (70% after 1.42% $R_{o,m}$), E increases to a
larger amplitude. With increasing X, so E reproduces the molecular structural characteristics of
coal so far as kinetics is concerned. As stated above, coal is a polymer with three-dimensional
network bearing a certain number of types and energy levels of chemical bonds, and the reaction
of hydrocarbon generated from coal degradation is the assemblage of a series of parallel and
continuous elementary reactions. With increasing temperature, when there is an increment ΔX
in the hydrocarbon generation rate X, it means the cracking of the chemical bonds of the
corresponding energy level. When X comes close to 1, the chemical bonds from low to high
energy are successively broken off, which shows the increase of activation energy in order. After
the hydrocarbon generation rate has reached 70%-80%, the content of aliphatic side chain, alicycle
and heteroatomic radicals from the hydrocarbon generated in coal is very low; the coal pyrolysis
at this time mainly destroys the aromatic C-C and C-H bonds, consequently it presents the
relatively high activation energy. In addition, as also known from Table 4, with increasing coal
rank, the apparent activation energy at the same rate of hydrocarbon generation displays on the
whole an increasing trend, but E varies relatively slow in the range of 0.58%-1.42%$R_{o,m}$; after the
reflectance of 1.42%, E varies with a larger amplitude, which is just in accordance with the second
coalification jump. Afterwards, the chemical reaction in coal transforms from hydrocarbon
generating to cracking, the hydrocarbons produced are split into smaller molecular gas
hydrocarbons. The aromatization degree of coal increases, and the aromatic rings are arranged in
order gradually due to the escape of volatile matter from coal. So the apparent activation energy of
coal displays the high values.

Relationship between the Distribution of Activation Energy and Coal Rank
From Figure 3a and Figure 3b, the distribution of activation energy is dispersive in the range of
0.42%-0.53% $R_{o,m}$, convergent in the medium metamorphic bituminous stage; afterwards, at the
high metamorphic bituminous stage, the distribution of activation energy is dispersive again.
Possibly it is because at the low coal rank, the side chain, branched chain and heteroatomic
radicals in coals are relatively more numerous, for there exists an assemblage of multienergy
levels of bonds, which results in the widespread distribution of activation energy; with increasing
coal rank, theoretically, the distribution of activation energy should turn to be convergent owing
to the continuous cracking and escape of side, branched chain and heteroatomic radicals in coals;
while the coal rank reaches to the high metamorphic bituminous stage, the radicals existing in coal,
such as side chain and branched chain that produced volatile matter are scarce, but the thermal
cracking of aromatic C-C and C-H bonds needs relatively high activation energy, so that it is in
general hard to be pyrolyzed completely at the temperature as high as about 800 °C[7]. As the
highest pyrolyzing temperature is only 600 °C in our experimental instrument, it results again in
the dispersive distribution of activation energy.

Geochemical Significance of Study on Reaction Kinetics of Coalification
It is generally thought that the hydrocarbon generation rate of 10%-90% is the hydrocarbon
window, the hydrocarbon generation rate of 10% is the threshold of hydrocarbon generation, 90%
is the lower limit of hydrocarbon generation. Judging from Table 3, both temperatures of
threshold and lower limit of hydrocarbon generation are relatively higher at the reflectance of
lower than 0.58% $R_{o,m}$ and   higher than 1.42% $R_{o,m}$, their corresponding temperature interval and

Liu Dameng et al.

the distribution interval of activation energy are relatively wide as well, which indicates the weak potential and wide scope of hydrocarbon generation. In the geological profile, the amount of hydrocarbon derived from coal is relatively small and dispersive with increasing depth or temperature, so it is difficult to form hydrocarbon pool.   Of course, essences of their hydrocarbon

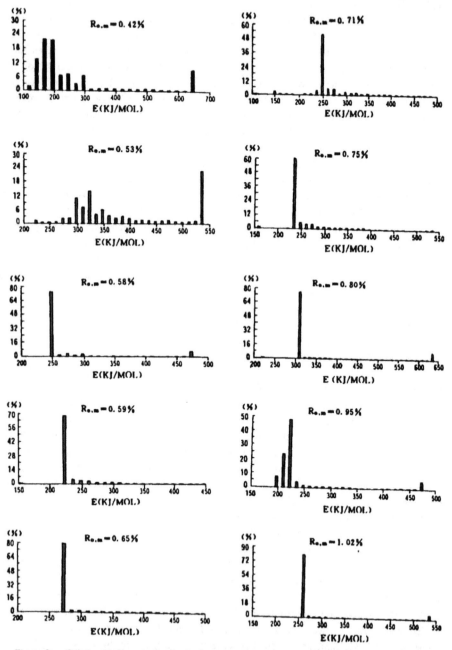

Figure 3a.   Relationship between the distribution of activation energy and hydrocarbon generation rate in various rank coals

generation are different. At the reflectance of lower than 0.58% $R_{o,m}$, the coal is under immature stage, the hydrocarbons derived from coal are mainly generated through biochemistry, and they contain large amount of biomarkers and heavy hydrocarbons; when the vitrinite reflectance is higher than 1.42% $R_{o,m}$, the hydrocarbons derived from coal are mainly cracked ones, containing large amount of light hydrocarbons, such as CH_4. The temperatures of threshold and lower limit of hydrocarbon generation are relatively low at the reflectance of 0.58%-1.42% $R_{o,m}$, their corresponding temperature intervals and activation energy intervals are relatively narrow, which shows a high potential and narrow scope of hydrocarbon generation at this rank stage. In the geological profile, one finds that a large amount of oil-gas will quickly be generated with increasing buried depth and/or temperature, resulting in the relative concentration of hydrocarbons, and the formation of oil-gas pool easier through migration and accumulation. The above-mentioned analysis of kinetics indicates that the hydrocarbons derived from coal may be enriched to form pool and they are mainly pyrolyzed hydrocarbons, while the cracked and biogenetic hydrocarbons are relatively little. Therefore, the searching for coal-derived hydrocarbons, should focus on exploring the medium-high mature areas to save manpower and material resources.

Figure 4 shows that the increasing of heating rate results in raising reaction temperatures with the same rate of hydrocarbon generation, the reaction temperatures at various heating rates all increase with increasing coal rank. It further proves that coalification of the same degree may result under relatively short-geological-term and high-temperature or relatively long-geological-time and low-temperature. Certainly, if the geotemperature is too low, the time

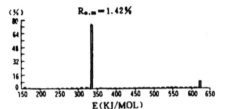

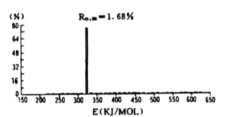

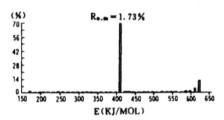

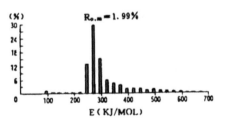

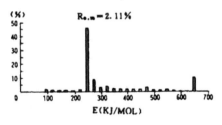

Figure 3b. Relationship between the distribution of activation energy and hydrocarbon generation rate in various rank coals

factor is of no effect. Under similar subsiding depth and geotemperature, if the duration of coalification is different, the coal rank formed is different as well; the longer the duration of heat undergoes, the higher the coalification degree is.

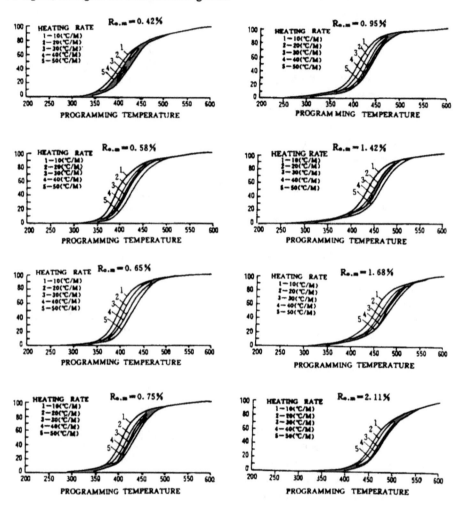

Figure 4. Relationship between pyrolyzing temperature and cumulative rate of hydrocarbon generation in various rank coals

The reaction rate increases with temperature at first; after it reaches the maximum value, the reaction rate decreases with increasing temperature(Fig.5). For the same rank coal, the increase of heating rate results in the moving of T_{max} value towards high temperature; as to the various rank coals, the T_{max} value increases with increasing coal rank at the same heating rate. Consequently, both heating rate and thermal evolution degree control the T_{max} value. In addition, according to the study of Burnham et al.[1], the T_{max} value varies with the pyrolyzed products, but their T_{max} values on the whole tend to be convergent with increasing coal rank. Therefore, when applying the result of pyrolysis analysis to evaluate the degree of thermal evolution, the components of samples used should be as close as possible, so that the comparisons of T_{max} values are of significance.

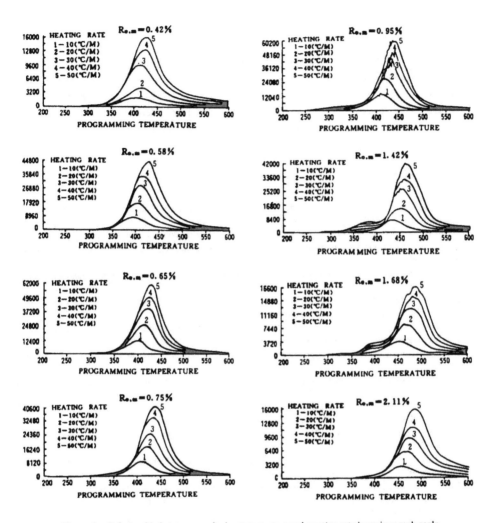

Figure 5. Relationship between pyrolyzing temperature and reaction rate in various rank coals

CONCLUSIONS

The study of reaction kinetics of coalification shows: (1) the transformation of molecular structure of coal is the main cause of coalification; (2) the hydrocarbons derived from coal will be enriched to form pool and they are mainly pyrolyzed hydrocarbons; the cracked and biogenetic hydrocarbons are relatively rare and dispersive, difficult to be enriched to form pool; (3) long-term and low-temperature or short-term and high-temperature can form the same rank coals; under similar temperature and subsiding depth, the longer the duration of heat experienced is, the higher the degree of coalification reaches; (4) the T_{max} value is not only related to heating rate and thermal evolution degree, but also to the component difference of source materials.

158   *Liu Dameng et al.*

Acknowledgments

This project is financially supported by the National Natural Science Foundation of China(NNSFC) Grant No.49172113. The authors would like to thank Prof. Yang Zunyi, Member of the Chinese Academy of Sciences for his helpful reviews of an early version of the manuscript. Thanks are due to Prof. Wu Liyan of Experimental Center, Beijing Research Institute of Petroleum Exploration and Development for fruitful and frequent discussions.

REFERENCES

1 Burnham,A.K., Oh, M S and Crawford, R W Pyrolysis of Argonne Premium coals: activation energy distributions and related chemistry, *Energy & Fuels* 3, 42-55(1989)
2 Hanbaba,P., Juntgen, H. and Peters,W Nichtisotherme reaktionskinetick der kohlenpyrolyse. Teil II: Erweiterung der theorie der Gasabspaltung und experimentalle bestatigung an steinkohlen, *Brennst. Chem* 49, 368-376(1968)
3 Hu Jianyi and Huang Difan *Theoretical bases of nonmarine petroleum geology in China*, Petroleum Industry Press,Biejing(1991).
4 Huang Difan, Li Jinchao and Zhang Dajiang. *Evolution and hydrocarbon generation mechanism of nonmarine organic matters*,Petroleum Industry Press, Beijing(1984).
5. Jin Qiang. Formation and hydrocarbon generation evolution of saline lake faces source rocks in Lower Tertiary of Dongpu depression, *East China Petroleum Institute Journal.*6, 10-16(1984).
6 Juntgen, H. and Van Heek, K.H. Gas release from coal as a function of the rate of heating, *Fuel* 48, 103-117(1968).
7 Juntgen, H Review of the kinetics of pyrolysis and hydropyrolysis in relation to the chemical construction of coal, *Fuel* 63, 731-737(1984).
8. Karweil,J Die metamorphse der kohlen vom standpunkt der physikalischen chemie, *Deutsch. Geol. Gesell. Zeitschr.* 107,132-139(1956)
9. Lou Shuangfang and Huang Difan Study of hydrocarbon generation kinetics of coal macerals, *Science in China Series B* 5, 812-817(1995).
10 Luo Qiuxia. Mathematic modelling of lower Tertiary source rock evolution in some basins of east China, *Petroleum and Experimental Geology* 2,19-23(1980)
11 Pitt,G J. The kinetics of the evolution of volatile products from coal, *Fuel* 41, 267-274(1962).
12 Reynolds, J.G and Burnham, A K. Pyrolysis kinetics and maturation of coals from the San Juan basin, *Energy & Fuels* 7, 610-619(1993).
13. Tissot,B.P , Pelet,R. and Ungerer,P. Thermal history of sedimentary basins, maturation indices and kinetics of oil and gas generation, *Bull.Am.Assoc.Pet.Geol.*71,1445-1446(1987).
14 Wang Daoyu and Wang Deijin. Numerical calculation of gross first-order reaction kinetics parameters of source rocks and oil shale pyrolysis, *East China Pet.Inst.J.*6,38-43(1984)
15. Wang Huixiang and Huang Difan. Study of pyrolysis kinetics of nonmarine kerogens, In: *Organic Geochem.and Nonmarine Pet. Formation*, Petroleum Industry Press, Beijing(1986).
16. Wang Jianqiu. Pyrolysis kinetics of hydrocarbon generation of petroleum source rocks using Rock-Eval, *East China Petroleum Inst. J.*6 (1984)
17. Wang Tingfen. Preliminary study of pyrolysis kinetics equation of oil shale in Fushun, *Petr.Proc.*2(1980).
18. Wang Tingfen. Pyrolysis kinetics of oil shale in Fushun using differential thermal analysis, *East China Pet.Inst.J.*3 (1981).
19. Wu Zhaoliang and Huang Xinhan. Calculation of oil and gas generation amount using kinetics model of hydrocarbon generation of source rocks, *East China Pet.Inst.J.*8(1986).
20. Yang Qi. Coalification.In:*Advances in Coal Geology*,Yang Qi(Eds), Science Press, Beijing(1987).
21 Yang Qi and Pan Zhigui. *Coal metamorphism and geological causes in Permo-Carboniferous, North China*, Geological Publishing House, Beijing(1988).
22. Yang Qi(Eds). *The Coal Metamorphism in China*, Coal Industry Publishing House, Beijing(1996).

23 Yang Wenkuan Approximate solution of first-order reaction equation and its application in quantitative predicting of gas and oil, *oil and natural Gas Geology* 3,99-112(1982)
24. Zhang Dajiang, Huang Difan and Li Jinchao. Pyrolysis kinetics of kerogen in oil shale and its geochemical significance, *Oil and Natural Gas Geology* 4, 383-393(1984).

Milton Keynes UK
Ingram Content Group UK Ltd.
UKHW040052071024
449327UK00019B/504